Nuclear Weapon Initiatives: Low-Yield R&D, Advanced Concepts, Earth Penetrators, Test Readiness

NUCLEAR WEAPON INITIATIVES: LOW-YIELD R&D, ADVANCED CONCEPTS, EARTH PENETRATORS, TEST READINESS

JONATHAN V. ERNEST
EDITOR

Novinka Books
New York

Editorial Production Manager: Donna Dennis
Senior Production Editor: Susan Boriotti
Production Editors: Marius Andronie, Rusudan Razmadze and Keti Datunashvili
Office Manager: Annette Hellinger
Graphics: Andrea Charles and Magdalena Nuñez
Editorial Production: Maya Columbus, Vladimir Klestov, Lorna Loperfido and Joanne Overton
Circulation: Ave Maria Gonzalez, Vera Popovic, Luis Aviles, Alexandra Columbus, Raymond Davis, Cathy DeGregory, Melissa Diaz, Marlene Nuñez and Jeannie Pappas
*
Library of Congress Cataloging-in-Publication Data
Nuclear weapon initiatives : low-yield R & D, advanced concepts, earth penetrators, test readiness / Jonathan V. Ernest, editor.
p. cm.
Includes index.
ISBN 1-59454-203-1 (softcover)
1. Nuclear weapons--United States. 2. United States--Military policy. 3. Military research--United States. 4. Nuclear weapons--Testing. I. Ernest, Jonathan V.
UA23.N7935 2004
355.02'17'0973--dc22 2004022332

Nova Science Publishers, Inc.
400 Oser Ave, Suite 1600
Hauppauge, New York 11788-3619
Tele.: 631-231-7269 Fax: 631-231-8175
e-mail: Novascience@earthlink.net
Web Site: http://www.novapublishers.com

Printed in the United States of America

CONTENTS

PREFACE

The Bush Administration completed its congressionally-mandated Nuclear Posture Review in December 2001. The review led to major changes in U.S. nuclear policy. It found that the Cold War relationship with Russia was "very inappropriate" and that this nation must be able to deal with new threats. It planned to retain Cold War-era nuclear weapons, which would suffice for many contingencies, though at reduced numbers. To complement these weapons so as to improve U.S. ability to deal with new, more dispersed threats in various countries, the Administration sought to explore additional nuclear capabilities.

Accordingly, the FY2004 request included four nuclear weapon initiatives: (1) rescinding the ban that Congress imposed in 1993 on R&D on low-yield nuclear weapons; (2) $6 million for the Advanced Concepts Initiative (ACI) to begin certain studies of weapon-related science and technology; (3) $15 million to continue a study of the Robust Nuclear Earth Penetrator (RNEP), in which an existing bomb would be converted into a weapon able to penetrate into the ground before detonating to improve its ability to destroy buried targets; and (4) $25 million to enable the United States to conduct a nuclear test within 18 months of a presidential order to test, and for related purposes, as compared with the current 24-36 month time that was set shortly after the end of the Cold War. Congress acted on these requests in the FY2004 National Defense Authorization Act (P.L. 108-136), and acted on the latter three in the FY2004 Energy and Water Development Appropriations Act (P.L. 108-137). For FY2005, the Administration requests $9.0 million for ACI, $27.6 million for RNEP, and $30.0 million for improving nuclear test readiness.

These initiatives are controversial. Supporters claim that the first three initiatives would enhance deterrence, thereby reducing the risk of war, and that some weapons that might result from the initiatives could enable the United States to destroy key targets in nations that may pose a threat. Critics are concerned that these initiatives would lead to nuclear testing, increase the risk of nuclear proliferation, and make U.S. use of nuclear weapons more likely. Regarding enhanced test readiness, the Administration argues that nuclear testing might be needed, for example, to check fixes to weapon types with defects, and that 24 to 36 months is too long to wait; critics are concerned that shortening the time to test could signal a U.S. intent to test, and that renewed testing could lead to a renewed interest in testing by other nations.

This book provides the policy context for the four initiatives. For each, it then presents a description, history, FY2004 legislative actions, the FY2005 request (for all but low-yield R&D), and issues for Congress. It is designed for those who want a detailed introduction to the debate, those seeking arguments and counterarguments, and those looking for answers to specific questions. It will track congressional and executive actions on these initiatives through updates as developments warrant.

Chapter 1

INTRODUCTION*

For FY2004, the Administration proposed four nuclear weapon initiatives that were small in terms of dollars and immediate impact, but that would have large policy ramifications and could lead to major outlays if implemented over time. Congress considered all of them in the FY2004 National Defense Authorization Act (NDAA) (P.L. 108-136), and considered all but the ban on R&D on low-yield weapons, which does not involve funding, in the FY2004 Energy and Water Development Appropriations Act (EWDAA) (P.L. 108-137) as well.[1] For FY2005, the low-yield R&D ban was not at issue, as Congress lifted it in the FY2004 NDAA; the Administration requested funds for the other three items below.

> *Rescission of a ban on R&D on low-yield nuclear weapons.* FY1994 legislation barred R&D that could lead to production of low-yield nuclear weapons. The Administration proposed rescinding this provision.
>
> *Advanced Concepts Initiative (ACI):* For FY2004, the Administration requested $15.0 million for the Robust Nuclear Earth Penetrator and $6.0 million for other ACI programs. The latter include computer modeling, remotely monitoring warheads and controlling their firing systems, and studying how to design warheads with specific radiation outputs and other effects. The FY2005 request is $27.6 million for the earth penetrator and $9.0 million for ACI.

---

* Excerpted from CRS Report RL32130 by Jonathan Medalia

[1] This report does not consider nuclear-armed missile defense interceptors. The Administration requested no funds for them for FY2004, and the Senate passed, by voice vote with no debate, an amendment by Senators Levin, Feinstein, and Stevens to bar use of FY2004 funds for work on such interceptors. U.S. Congress. *Congressional Record,* May 21, 2003: S6836.

Robust Nuclear Earth Penetrator (RNEP): The request would continue a study of modifying an existing weapon to penetrate completely into the ground before detonating, increasing its ability to destroy buried targets.

Nuclear Test Readiness: A "test readiness posture" is the maximum time between a presidential order to conduct a nuclear test and the test itself. Since 1996, the U.S. posture has been 24 to 36 months. For FY2004, the Administration requested $24.9 million to maintain this posture (for which at least $15.0 million is needed) and to begin changing it to 18 months. The FY2005 request is $30.0 million.

These initiatives are controversial. Many on both sides of the issue see them as a gateway to new weapons. They collectively give new impetus to the decades-long debate over the role of nuclear weapons in promoting U.S. security.

The four initiatives will likely be at issue for some years to come because they could lead to increased requests in the future. While it costs nothing to lift the ban on low-yield R&D, doing so would presumably lead to low-yield R&D that would have a cost. The FY2004 request for ACI was $6.0 million, but there were calls to increase its funding substantially. The FY2004 RNEP request was $15.0 million to continue a study, but after the study ends in FY2006, the request would grow sharply if Congress and the Administration decide to develop the weapon. NNSA estimates that it would cost $83 million over three years to implement an 18-month test readiness posture, and $25 million to $30 million a year to sustain it thereafter, compared to the $10 million or so spent each year to maintain the current 24- to 36-month posture.[2]

This book discusses these initiatives and the issues they raise. This section presents background and key arguments; subsequent sections provide a history and issues for Congress for each initiative. The book is designed for those who want a detailed introduction to the debate, including its history; those seeking arguments and counterarguments; and those looking for answers to specific questions. It will track congressional and executive actions on these initiatives through regular updates as developments warrant.

---

[2] U.S. Department of Energy. National Nuclear Security Administration. *Report to Congress: Nuclear Test Readiness,* April 2003, p. 1.

Chapter 2

THE BROADER CONTEXT FOR THE FOUR INITIATIVES

During the Cold War, the focus of U.S. nuclear policy was deterring or retaliating in response to a Soviet nuclear attack. U.S. nuclear forces were designed and sized for that contingency. Other contingencies in which the United States might have used nuclear weapons were much more remote, and would have required far fewer nuclear weapons, than responding to a Soviet attack. Defense planners dealt with them, but as "lesser included cases" not requiring new weapons.

The end of the Cold War led to a reduction of the U.S. nuclear weapons development effort. The first Bush Administration halted nuclear testing except for safety and reliability of the nuclear stockpile, canceled weapons that were under development, and withdrew most U.S. battlefield nuclear weapons. Numbers of personnel in the nuclear weapons complex — the facilities managed by the Department of Energy (DOE) to develop, test, manufacture, and maintain nuclear weapons — dropped sharply. At congressional direction, the United States began a moratorium on nuclear testing in October 1992 that is still in effect and worked to negotiate the Comprehensive Test Ban Treaty (CTBT), a ban on all nuclear explosions. Before the moratorium, the United States conducted nuclear tests on an ongoing basis; with the end of testing, the Clinton Administration changed the test readiness posture to 24 to 36 months.

The George W. Bush Administration came into office with a different stance on nuclear weapons policy, which the September 11 attacks stiffened. It delivered a Nuclear Posture Review (NPR) to Congress in December 2001.[1]

---

[1] For further information on U.S. nuclear policy, see CRS Report RL31623, U.S. Nuclear Weapons: Changes in Policy and Force Structure, October 28, 2002, 44 p., and CRS Report

This classified study, mandated by legislation,[2] spelled out the Administration's views on the role of nuclear weapons. A strategic relationship with Russia based on the threat of mutual annihilation had become "very inappropriate,"[3] according to J.D. Crouch, Assistant Secretary of Defense for International Security Policy, and the threat had become more diffuse. The NPR focused on nuclear capabilities deemed needed for various contingencies rather than mainly to counter the Soviet nuclear threat. That is, the Administration perceived a wider range of possible uses for nuclear weapons quite different than the U.S.-Soviet nuclear war scenarios that dominated Cold War strategy. Crouch said, "The capabilities-based approach argues that there may be multiple contingencies and new threats that we have to deal with. We're focusing on how we will fight ... not who or when."[4] To support this approach, the NPR envisioned a "new triad" — strike forces (nuclear and nonnuclear); defenses; and an infrastructure better able to respond to military needs — tied together by command and control, intelligence, and planning. The goal was to "assure allies and friends" that the United States had "credible non-nuclear and nuclear response options," "dissuade competitors" through a "diverse portfolio of capabilities [that] denies payoff from competition," "deter aggressors," and "defeat enemies."[5]

The NPR called for retaining the capability to deter or respond to a variety of contingencies, and for retaining Cold War-era nuclear weapons, at lower levels, to do so. With the demise of the Soviet Union and the September 11 attacks, a subset of targets — terrorist-related facilities, some of which had existed for decades — assumed greater prominence. One example is hardened and deeply buried facilities. Rogue (and many other) states have built such facilities, which might house leaders and key communications facilities. Another example is facilities — some deeply buried, others not — for producing or storing weapons of mass destruction (WMD).

The end of the Cold War also led to changed constraints on use of nuclear weapons. In a nuclear war with the Soviet Union, assuring the immediate

RS21619, Nuclear Weapons and U.S. National Security: A Need for New Weapons Programs?, September 15, 2003, 6 p., both by Amy Woolf.

2 The requirement for the NPR is in P.L. 106-398, Floyd D. Spence National Defense Authorization Act for Fiscal Year 2001, Section 1041.

3 U.S. Department of Defense. "Special Briefing on the Nuclear Posture Review." Presented by J.D. Crouch, Assistant Secretary of Defense for International Security Policy, January 9, 2002. [http://www.defenselink.mil/news/Jan2002/t01092002_t0109npr.html]

4 Ibid.

5 U.S. Department of Defense. "Findings of the Nuclear Posture Review," briefing slides, January 9, 2002, slide titled "QDR: Defense Policy Goals," available at [http://www.defenselink.mil/news/Jan2002/g020109-D-6570C.html], and U.S. Department of Defense. "Special Briefing on the Nuclear Posture Review," January 9, 2002.

destruction of thousands of targets of all sorts would have been critical to national survival.

Reducing fallout and "collateral damage," or damage to people and things that were not the intended targets, was secondary in importance to achieving required probabilities of damage on targets. Warheads were typically designed to achieve that primary goal. Now, nuclear weapons to hold at risk specific, limited targets in rogue states in a credible manner are judged to need several characteristics:

- They would need to be able to counter threats that rogue states might pose, such as hard and deeply buried facilities or production facilities for WMD. Weapons to attack buried facilities would have different characteristics than weapons to defeat chemical and biological agents.
- They would be better able to minimize (*not* eliminate) collateral damage consistent with military objectives. While this has long been a goal of U.S. nuclear policy, earth penetration and increased accuracy sharply reduce the yield needed to destroy a target, making it more feasible to attain that goal.
- They would need to be available readily, if not immediately, because the specific threat might become known with little warning.
- They could be few in number. While an all-out nuclear attack on the Soviet Union would have struck thousands of targets, the option of an attack on a rogue state might involve a handful of targets.

According to the National Nuclear Security Administration (NNSA),[6] "The Administration believes the broader range of capabilities of a nuclear stockpile with these weapons will serve as a more credible, and hence more effective, deterrent than the Cold War stockpile we have today. This more effective deterrent will make the use of nuclear weapons less likely."[7]

Some critics believe that the only use for nuclear weapons should be to deter and, if necessary, respond to the use of nuclear weapons against the United

6 NNSA is a semiautonomous agency within the Department of Energy (DOE) that has managed the nuclear weapons complex since 2001. This complex is composed of laboratories, manufacturing plants, and a test site that design, develop, manufacture, maintain, and dismantle nuclear weapons, and that also used to conduct nuclear explosions to test them. (Before 2001, DOE managed this complex directly through its Office of Defense Programs.)

7 Comments provided to the author by the National Nuclear Security Administration, August 18, 2003.

States.[8] They oppose attempts to develop nuclear weapons with lower yield and reduced collateral damage as blurring the distinction between conventional and nuclear weapons. Even U.S. use of a low-yield atomic bomb, they assert, would result in a firestorm of protest and worldwide opprobrium.

Further, they believe that the U.S. nuclear arsenal has long been sufficient against a range of threats. Hardened and deeply buried targets (HDBTs), for example, have existed for decades, yet the first Bush Administration halted work on nuclear weapons then under development and halted nuclear testing except for safety and reliability, effectively bringing work on new weapon types to a close. Presumably that administration would not have taken these steps had it envisioned a need for new weapon types.

The United States needs the active support of the entire international community in the war against terrorism, many observers believe. Assistance may take many forms: providing information to the CIA on suspicious individuals, safeguarding shipping containers, securing radioactive material, or closing bank accounts of suspected terrorists. Other nations will be more likely to cooperate on issues important to the United States, it is argued, if the United States cooperates with other nations on issues important to them. One such issue is nuclear nonproliferation. The United States pledged in the Nuclear Nonproliferation Treaty (Article VI) "to pursue negotiations in good faith on effective measures relating to cessation of the nuclear arms race at an early date and to nuclear disarmament." In addition, as an inducement in securing international support for extending the treaty indefinitely, the United States and other nuclear weapon states pledged in 1995 to support the CTBT.[9] They made a similar statement at the 2000 NPT review conference.[10]

Critics maintain that the Administration's nuclear initiatives are at odds with these pledges. They argue that developing new or modified weapons runs counter to the NPT pledge and that such weapons will ultimately require nuclear testing to ensure that they work. Reducing the time to conduct a nuclear test supports a future decision to test and, to the critics, implies an intent to test.

---

8 See, for example, Union of Concerned Scientists, "Global Security: Nuclear Weapons," [http://www.ucsusa.org/global_security/nuclear_weapons/index.cfm?pageID=30].

9 For further information on this treaty, including subsequent Senate action, see CRS Issue Brief 92099, *Nuclear Weapons: Comprehensive Test Ban Treaty,* CRS Report 97-1007, *Nuclear Testing and Comprehensive Test Ban: Chronology Starting September 1992,* and CRS Report RS20351, *Comprehensive Test Ban Treaty: Pro and Con,* all by Jonathan Medalia.

10 Statement by the Delegations of France, the People's Republic of China, the Russian Federation, the United Kingdom of Great Britain and Northern Ireland, and the United States of America to the 2000 NPT Review Conference, May 1, 2000. Available at: [http://www.ceip.org/files/projects/npp/resources/npt2000p5.htm].

These positions may harm U.S. security by undermining efforts to make worldwide cooperation on nonproliferation of nuclear and other WMD more effective. Even if some steps might be innocuous by themselves, critics believe that many around the world will see the confluence of them as pieces of a single policy and will infer that the United States is now much more willing to consider using nuclear weapons. As an example of international concern over the new U.S. policies, the foreign ministers of the New Agenda Coalition (Brazil, Egypt, Ireland, Mexico, New Zealand, South Africa, and Sweden) issued a statement that said in part,

> 4. The Ministers stressed that each article of the NPT is binding on the respective States parties, at all times and in all circumstances ...
>
> 7. The Ministers reiterated their deep concern at emerging approaches to the broader role of nuclear weapons as part of security strategies, including rationalizations for the use of, and the development of new types of nuclear weapons.[11]

These positions may undermine U.S. security, critics fear, by spurring nations that might be targets of a U.S. nuclear attack to step up their efforts to acquire nuclear weapons. Of the three "Axis of Evil" nations, Iraq, the first target of a U.S. attack, did not have nuclear weapons as far as is currently known. Having nuclear weapons to deter the United States may be a consideration for North Korea and Iran, both of which apparently have active programs to produce fissile material. North Korea has stated that it has several nuclear weapons; while Iran has declared that its nuclear program is for peaceful purposes, few in the United States believe that assertion.

Even in military terms, critics argue, nuclear use would offer marginal to negative value. Deeply buried targets could be defeated (not necessarily destroyed) with conventional weapons, destruction of entrances and air shafts, or demolition by special forces. The effectiveness of low-yield nuclear weapons for destroying biological or chemical agents is uncertain, as discussed below, and intelligence might be unable to locate targets in underground complexes, as recent efforts to find Iraqi WMD showed, rendering nuclear weapons of any yield of little use. Nations could counter earth penetrators by burying facilities more deeply. Even low-yield or earth penetrating weapons would throw a large amount of radioactive debris into the atmosphere. Substituting radioactive

---

[11] "Deep Concern At The lack Of Progress," declaration, issued by the Foreign Ministers of the New Agenda Coalition, United Nations Headquarters, New York, 23 September 2003. [http://www.acronym.org.uk/docs/0309/doc20.htm]

materials for biological or chemical agents may offer little advantage to the local population or U.S. troops.

Chapter 3

R&D ON LOW-YIELD NUCLEAR WEAPONS

DESCRIPTION

A provision in the FY1994 National Defense Authorization Act barred R&D that could lead to production by the United States of a nuclear weapon of less than 5 kilotons (kt)[1] that had not entered production by November 1993. In its FY2004 legislative proposals, the Administration asked Congress to rescind this provision on grounds that it undercut U.S. ability both to explore technical options that could deter or respond to emerging threats, and to revitalize the nuclear weapons enterprise. This provision is often referred to as the Spratt-Furse provision, after its legislative authors, Representatives John Spratt, Jr., and Elizabeth Furse, or as the PLYWD provision, for Precision Low-Yield Weapon Design. The Armed Services Committees considered the proposed rescission; the Appropriations Committees did not do so because it involved no money. The FY2004 NDAA, Section 3116, repealed the Spratt-Furse provision but (1) made clear that the repeal did not "authoriz[e] the testing, acquisition, or deployment of a low-yield nuclear weapon" and (2) barred the Secretary of Energy from beginning engineering development or subsequent phases of work on a low-yield weapon without specific congressional authorization. With the provision lifted, there is no parallel request for FY2005.

---

1 One kiloton is equivalent to the explosive force of 1,000 tons of TNT; the yield of the Hiroshima bomb was 15 kilotons.

TECHNICAL BACKGROUND

Military Utility of 5-Kt Weapons

The debate over the Spratt-Furse provision treated 5 kt as critical because that was the threshold in the law. Those who would lift the ban argued that the ban made it difficult to train weapon designers and impossible to develop essential new weapons; their opponents were concerned that lifting the ban would signal U.S. interest in developing capabilities that would make the use of nuclear weapons more likely. Yet the threshold was of only slight relevance to the military or the weapons labs because very low yield nuclear weapons have uncertain military utility. Substantial advances would be needed in the understanding of low-yield weapon effects to assess the military utility and collateral damage of such weapons.

Military utility derives from nuclear-weapon effects that are prompt and that can be measured with reasonable precision. Damage from the blast wave falls into this category; fallout and secondary fires (e.g., fuel set afire) do not. For a 500-kt weapon, the blast effects are highly predictable: almost anything on the Earth's surface over a large, known area will be destroyed. (For comparison, the Hiroshima bomb had a yield of 15 kt.) At the other end of the scale, the blast effects of a 500-pound conventional bomb are strongly influenced by accuracy, height of burst, local features, etc. For example, a hill or building can block or redirect the blast wave. Such factors also shape the effects of a low-yield nuclear weapon, and their influence increases as yield decreases. Modeling such factors is more complicated than for high-yield weapons, has been done to a much lesser extent, and depends more on unknown details such as a building's construction. It may be that detailed engineering studies and computer models of the prompt effects of existing low-yield nuclear weapons could provide as much military value as designing new low-yield weapons.

A 1979 report by the Office of Technology Assessment elaborated on the problem of uncertain weapon effects in discussing a 1-kt terrorist nuclear weapon detonated at ground level in a city:

> the highly built-up urban structure in which the weapon is placed will significantly modify the resulting nuclear environment. This occurs when the lethal range of effects shrink to such an extent that they are comparable to the size of urban structures. It is indeed reasonable to expect that the blast effects of a small weapon ... will be severely influenced by nearby structures having comparable dimensions. Preliminary calculations have confirmed this. ... In summary, the ranges of nuclear effects from a low-yield explosion in the

confined space of an urban environment will differ significantly from large yield effects, but in ways that are very difficult to estimate.[2]

If the Warsaw Pact had invaded Western Europe during the Cold War, such uncertainties of low-yield weapons would have been immaterial for some missions (an atomic demolition munition could destroy a bridge, and an artillery-fired atomic projectile could destroy troop concentrations, more reliably than conventional munitions) and in some situations (as a last resort to keep front-line troops from being overrun). Now, though, extreme accuracy may enable conventional weapons to destroy small targets, and massive conventional bombing can attack troop concentrations. At the same time, this same accuracy may give low-yield nuclear weapons new military missions by enabling them to destroy targets that in the past would have required much larger weapons. On the other hand, low-yield weapons face uncertainties that render them of questionable value in going after key targets, such as buried facilities for which details of the structure and its overlying earth and rock are poorly known. This uncertainty increases as yield drops. For reasons such as these, George Miller, formerly Associate Director for Defense and Nuclear Technologies at Lawrence Livermore National Laboratory, observed, "The vast majority of the 'low yield' concepts that would provide substantial reductions in collateral damage have yields greater than 5 kt. Concepts in the less than 5 kt regime are at an early R&D stage."[3]

History through the FY2003 Budget Cycle

The United States has had many low-yield weapons in the stockpile for decades.[4] In the wake of the 1991 Persian Gulf War, the U.S. military became more interested in them for attacking underground structures. One news report stated that "[p]enetrating Saddam's hardened bunkers proved to be one of the most daunting tasks that faced the Air Force."[5] In late 1991, the Air Force

2 U.S. Congress. Office of Technology Assessment. *The Effects of Nuclear War.* OTA-NS-89, USGPO, May 1979, p. 45-46.

3 Personal communication, August 15, 2003.

4 Thomas Dowler and Joseph Howard III, "Countering the Threat of the Well-Armed Tyrant: A Modest Proposal for Small Nuclear Weapons," *Strategic Review,* Fall 1991, "Countering the Threat of the Well-Armed Tyrant," p. 34-40, discuss low-yield nuclear weapons and their military advantages. William Arkin, "Nuclear Junkies: Those Lovable Little Bombs," *Bulletin of the Atomic Scientists,* July-August 1993: 22-27, discusses the bureaucratic progress of such weapons and their drawbacks.

5 Tony Capaccio, "Mini-Nukes Weighed for Third World Contingencies," *Defense Week,* September 8, 1992: 12.

reportedly asked Los Alamos National Laboratory to study a very low yield warhead that could destroy underground bunkers.[6] A 1992 Lawrence Livermore National Laboratory journal stated, "Several possibilities for new types of weapons are being evaluated for prototype development. These include ... low-yield earth-penetrating warheads for attacking underground command posts with minimum collateral damage..."[7] DOE began a concept definition study for an Aircraft Delivered Precision Low-Yield Weapon in FY1992, which it planned to continue in FY1993 and FY1994.[8]

At the same time, many in Congress and elsewhere sought to restrain U.S. nuclear weapons development. Legislation mandated a U.S. nuclear testing moratorium, which started in October 1992 and continues to the present, and directed the President to provide a plan for achieving a multilateral CTBT by September 30, 1996.[9] Preparations were underway in the U.N. in 1993 to begin CTBT negotiations in 1994. The NPT review and extension conference, which would decide whether to extend the treaty indefinitely, was planned for April 1995. Some saw testing, and weapons development that could lead to testing, as unhelpful to these efforts.

In 1993, the House Armed Services Committee, concerned over efforts by DOE to study low-yield nuclear weapons, included a provision barring R&D on these weapons in its version of the FY1994 National Defense Authorization Act, H.R. 4201. The committee's report stated:

> The committee is aware of recent efforts by the department [of Energy] to perform concept and feasibility studies for designing very low yield nuclear weapons. The committee opposes these efforts. Very low yield nuclear warheads threaten to blur the distinction between conventional and nuclear conflict, and could thus increase the chances of nuclear weapons use by another nation. In addition, the committee believes that the development of very low yield nuclear weapons undermines U.S. efforts to discourage nuclear weapons development by other nations, and would undercut U.S. efforts to negotiate an extension of the Non-Proliferation Treaty or a Comprehensive Test Ban. Finally, the utility of very low yield nuclear weapons is questionable given the increasing effectiveness and availability of precision guided conventional munitions.

---

6 "Lab Proposes Smaller Nuclear Weapons," Associated Press (newswire), October 8, 1992.

7 "Future Directions for the Nuclear Weapons Program at LLNL [Lawrence Livermore National Laboratory]," *Energy and Technology Review,* January-February 1992: 18.

8 U.S. Department of Energy. Office of Chief Financial Officer. *FY1994 Congressional Budget Request, Volume 1: Atomic Energy Defense Activities.* DOE/CR-0012, April 1993, p. 45-46.

9 Sec. 507 of P.L. 102-377, Energy and Water Development Appropriations Act, FY1993, contains these requirements.

> The committee therefore recommends a provision (sec. 3105 (e)) that would direct the Secretary of Energy to discontinue the ongoing concept design work within the department's nuclear weapons laboratories and to refrain from any future feasibility, engineering, development, or production work associated with very low yield nuclear weapons. The committee further directs the Secretary to work with the President, and other interested agencies in discouraging the development of similar weapons in other countries.[10]

The conference report summarized the House position, noted that the Senate bill had no similar provision, and stated:

> The Senate recedes with an amendment that would clarify that the prohibition applies to activities that could lead to production of new low-yield weapons. While the conferees agree that the provision is intended to prohibit research and development geared toward the production of any low-yield nuclear weapons by the United States, the conferees recognize that there are instances where the Department of Energy may need to conduct research on these types of weapons for other purposes. This would include research, in the interest of counter-proliferation, on the designs of low-yield weapons as a way to: (1) understand others' activities, including potential terrorist threats; (2) provide information for export control activities; and (3) understand the potential damage that could be inflicted by the use of these types of weapons. In addition, the conferees agree that nothing in this section would prohibit the Department of Energy from performing the research and development necessary for modifications to existing weapons in order to address safety or reliability problems.

The conferees direct the Secretary to work with the President and interested agencies in discouraging the development of similar weapons in other countries.[11]

The final legislation (P.L. 103-160, November 30, 1993, 107 stat 1946) follows:

SEC. 3136. PROHIBITION ON RESEARCH AND DEVELOPMENT OF LOW-YIELD NUCLEAR WEAPONS.

UNITED STATES POLICY- It shall be the policy of the United States not to conduct research and development which could lead to the production by the United States of a new low-yield nuclear weapon, including a precision low-yield warhead.

---

[10] U.S. Congress. House. Committee on Armed Services. *National Defense Authorization Act for Fiscal Year 1994.* H.Rept. 103-200, USGPO, 1993, p. 427.

[11] U.S. Congress. Committee of Conference. *National Defense Authorization Act for Fiscal Year 1994.* H.Rept. 103-357, USGPO, 1993, p. 840-841.

LIMITATION- The Secretary of Energy may not conduct, or provide for the conduct of, research and development which could lead to the production by the United States of a low-yield nuclear weapon which, as of the date of the enactment of this Act, has not entered production.

EFFECT ON OTHER RESEARCH AND DEVELOPMENT- Nothing in this section shall prohibit the Secretary of Energy from conducting, or providing for the conduct of, research and development necessary —

to design a testing device that has a yield of less than five kilotons;
to modify an existing weapon for the purpose of addressing safety and reliability concerns; or
to address proliferation concerns.

DEFINITION- In this section, the term `low-yield nuclear weapon' means a nuclear weapon that has a yield of less than five kilotons.

Several years later, the Office of Defense Programs asked DOE's General Counsel about the legal meaning of this prohibition. In a memorandum of March 1999, she concluded as follows:

> Section 3136 would appear to prohibit the Department from taking or supporting any action that could result in producing a weapon with a yield of less than 5 kilotons, unless one of the following four exemptions applies: (1) the activities are related to a weapon that entered production prior to December , 1993; (2) the activities are supporting design of a testing device; (3) the activities are directed to modifying an existing weapon to address safety and reliability concerns; or (4) the activities are supporting counterproliferation work.
>
> As you can see, the question of legality on any course of action involving low-yield nuclear weapons is highly fact dependent.[12]

In its report on the FY2001 defense authorization bill, S. 2549, the Senate Armed Services Committee made clear the issue prompting DOE's request to the General Counsel and sought to redress that issue legislatively:

> The committee recommends a provision that would require the Secretaries of Defense and Energy to assess requirements and options for defeating hardened and deeply buried targets. The provision would expressly authorize the

[12] Memorandum from Mary Anne Sullivan, General Counsel, to Thomas F. Gioconda, Brigadier General, USAF, Principal Deputy Assistant Secretary for Military Applications, Office of Defense Programs, "Statutory Prohibition on Research and Development of Low-Yield Nuclear Weapons," March 19, 1999.

> Department of Energy (DOE) to conduct any limited research and development that may be necessary to complete such assessments.
>
> The committee notes that a recent legal interpretation of existing law raised questions regarding whether DOE could participate in or otherwise support certain Department of Defense (DOD) studies and options assessments for defeating hardened and deeply buried targets. This provision removes any uncertainty and expressly allows DOE to assist the DOD with a review of these targets and the options for defeating such targets. The committee believes that DOE should provide information and all other assistance required to help DOD make informed decisions on whether: (1) to proceed with a new method of defeating hardened and deeply buried targets and; (2) to seek any necessary modifications to existing law.
>
> The committee is concerned that the ability to defeat hardened and deeply buried targets will continue to be a significant challenge for the foreseeable future.[13]

Although the House bill did not include a similar provision, the conference bill, H.R. 4205, did. Section 1044, "Report on the defeat of hardened and deeply buried targets," required the Secretary of Defense, in conjunction with the Secretary of Energy, to conduct a study on this topic, due by July 1, 2001.[14] That section directed that the study review U.S. requirements to defeat such targets "and stockpiles of chemical and biological agents and related capabilities," review plans to meet the requirements and determine the adequacy of the plans, "identify potential future hardened and deeply buried targets," determine what is needed to defeat them and to defeat stockpiles of chemical and biological agents, assess options to defeat such targets, and "determine the capability and cost of each option..." Section 1044 gave the Secretaries authority to "conduct any limited research and development that may be necessary to perform those assessments." Further, "The conferees believe that DOE should provide information and other assistance required to help DOD make informed decisions on whether: (1) to proceed with a new method of defeating hardened and deeply buried targets; and (2) to seek any necessary modifications to existing law."[15]

---

[13] U.S. Congress. Senate. Committee on Armed Services. *National Defense Authorization Act for Fiscal Year 2001.* S. Rept. 106-292. 106th Congress, 2nd Session. USGPO, 2000, p. 348-349.

[14] For text of Section 1044, see U.S. Congress. Committee of Conference. *Enactment of Provisions of H.R. 5408, the Floyd D. Spence National Defense Authorization Act for Fiscal Year 2001.* H.Rept. 106-945. 106th Congress, 2nd Session. USGPO, 2000, p. 277-278.

[15] Ibid., p. 851. The President signed the measure into law on October 30, 2000 (P.L. 106-398).

The DOD-DOE report required by Section 1044, dated July 2001, contained a rationale for low-yield nuclear earth penetrator weapons.[16] It stressed the problem that hardened and deeply buried targets (HDBTs) posed to the United States:

> Our potential adversary's weapons of mass destruction (WMD), long-range missiles, modern air defenses, most sophisticated command and control systems, national leadership in wartime, and a variety of tactical arms are increasingly concealed and protected by networks of hard and deeply buried facilities. If the United States does not have the means to defeat these facilities and the threatening assets they protect, adversaries may perceive that they have a sanctuary from which to coerce or attack the United States, its allies, or its coalition partners with threats much more powerful than in past conflicts. (p. 3)

It discussed the potential problem posed by buried facilities containing chemical or biological weapons (CBW):

> Ordnance employing fragmentation and blast effects will not accomplish this objective [destroying agents within munitions or containers] and may further worsen the situation by releasing agents into the atmosphere and surrounding environment. In some situations, there may be a need for multiple types of payloads to accomplish several objectives. For example, in the case of CBW located within a hardened facility, the goals might be *in situ* neutralization of the agents plus access-denial that prevents adversaries from recovering and using agents or production equipment not destroyed. This class of problems is the most vexing challenge to defeat of HDBTs. (p. 14)

While the DOD-DOE report focused on conventional weapons, special operations forces, intelligence, etc., for defeating HDBTs, it also discussed the potential value of nuclear weapons for this mission, especially low-yield penetrators:

> DoD and DOE have completed initial studies on how existing nuclear weapons can be modified to defeat those HDBTs that cannot be held at risk with conventional high-explosive weapons or current nuclear weapons. (p. 4)

> ... we also must prepare for those unique and emerging strategic threats that are critical and well protected ... This will require additional investment in intelligence, special weapons, and counter-WMD capabilities, including nuclear weapons. (p. 6)

---

[16] U.S. Departments of Defense and Energy. Report to Congress on the Defeat of Hard and Deeply Buried Targets, Submitted by the Secretary of Defense in Conjunction with the Secretary of Energy in Response to Section 1044 of the Floyd D. Spence National Defense Authorization Act for Fiscal Year 2001, PL 106-398, July 2001, 24 p.

> Nuclear weapons have a unique ability to destroy both agent containers and CBW agents. Lethality is optimized if the fireball is proximate to the target. This requires high accuracy; for buried targets, it also may require a penetrating weapon system. Given improved accuracy and the ability to penetrate the material layers overlying a facility, it is possible to employ a much lower-yield weapon to achieve the needed neutralization. The ability to use a lower yield would reduce weapon-produced collateral effects. The current nuclear weapons stockpile, while possessing some limited ground penetration capability and lower yield options (not yet certified), was not developed with this mission in mind. (p. 19)

Despite the above, the DOD-DOE report ended by noting, "DoD has not defined a requirement for a nuclear weapon for WMD Agent Defeat missions." (p. 24)

The FY2002 National Defense Authorization Act, and its predecessor versions in House and Senate, did not address the issue of low-yield nuclear weapons or nuclear earth penetrators.

As noted, the Administration completed its congressionally-mandated Nuclear Posture Review (NPR) in December 2001. A text purported to consist of leaked excerpts of the NPR surfaced in March 2002 and was widely quoted in the press.[17] Among other things, this text stated that, compared to a surface-burst nuclear weapon of a given yield, a nuclear earth penetrator weapon could destroy many buried facilities with much lower yield and thereby reduce fallout by a factor of 10 to 20.[18]

In action on the FY2003 National Defense Authorization Act, Representative Weldon initially offered an amendment on the House floor that, among other things, would have repealed the Spratt-Furse provision upon presidential certification that another nation conducted a nuclear test for new or improved nuclear weapons, or that another nation was developing, in underground facilities, WMD that could pose an imminent risk to the United States, or that the repeal "is in the national security interest of the United States."[19] Before the amendment was debated, Representatives Spratt and

---

[17] *Nuclear Posture Review [Excerpts],* available at [http://www.globalsecurity.org/wmd/library/policy/dod/npr.htm]. This is not an official source. Secretary of Defense Donald Rumsfeld confirmed that the NPR was leaked. When asked about the leak at a press conference, he said, "I am disturbed by the leak because I think it's enormously unprofessional." U.S. Department of Defense. News Transcript: "DoD News Briefing — Secretary Rumsfeld and Gen. Pace," March 15, 2002. [http://www.defenselink.mil/news/Mar2002/t03152002_t0315sd.html]. Michael Gordon, "U.S. Nuclear Plan Sees New Targets and New Weapons," *New York Times,* March 10, 2002: 1, 6, among others, draws on leaked excerpts of the NPR.

[18] Nuclear Posture Review [Excerpts], p. 47.

[19] U.S. Congress. *Congressional Record,* May 9, 2002: H2334.

Weldon subsequently modified it. While the 1993 provision barred "research and development which could lead to the production by the United States of a new low-yield nuclear weapon," the modification barred such development but not research, and clarified "development" as follows: "The term 'development' does not include concept definition studies, feasibility studies, or detailed engineering design work."[20] The amendment passed, 362-53. The Senate bill did not have a similar provision and the House provision was dropped in conference.

A description of the "phases" of weapon development is needed to understand the subsequent discussion of legislation on low-yield R&D and the other three nuclear initiatives. The development of new nuclear weapons proceeds through a series of defined "phases." DOD and the Atomic Energy Commission, a predecessor of DOE, entered an agreement on March 21, 1953, defining these phases as follows: Phase 1, weapon conception; phase 2, program study; phase 2a, design definition and cost study; phase 3, development engineering; phase 4, production engineering; phase 5, first production; phase 6, quantity production and stockpile; and phase 7, retirement.[21] (Note that phase 2 involves some prototype and subsystem testing as well as paper and computer studies.) The key dividing line is between phase 2a (design definition) and phase 3 (full-scale development). In phase 2a, Los Alamos and Lawrence Livermore National Laboratories, the two laboratories that design the nuclear explosive components of nuclear weapons, would generate one or more designs each for a particular warhead; personnel from Sandia National Laboratories, which design nonnuclear components of a warhead, would participate in the two design teams. At the end of phase 2a, one option and laboratory would be selected. (No new warhead has been designed since the 1980s.) In phases 3 and 4, which now would occur concurrently, the phase 2a design would be turned into a detailed design for a producible weapon through computer simulation, nuclear and nonnuclear testing, etc. For alterations and modifications of existing weapons, those in phase 6, the phase structure above is repeated as the "X" in "phase 6.X," though with some different names: phase 6.1, concept assessment; phase 6.2, feasibility study and option down-select; phase 6.2a, design definition and cost study; phase 3, development engineering; phase 4, production engineering; phase 6.5, first production; and phase 6.6, full-scale production.[22]

---

[20] Ibid., p. H2335.

[21] U.S. Atomic Energy Commission. "An Agreement Between the AEC and the DOD for the Development, Production, and Standardization of Atomic Weapons," March 21, 1953, 10 p. Phases 2a and 7 were added later.

[22] U.S. Department of Defense, Department of Energy, and Nuclear Weapons Council. "Procedural Guideline For The Phase 6.X Process," April 19, 2000, 13 p., defines these phases and explains the roles of DOD and DOE in weapons development.

With that as background, Section 3143 of P.L. 107-314, FY2003 National Defense Authorization Act, "Requirements for Specific Request for New or Modified Nuclear Weapons," required DOE (1) to have a single line item for all funds requested for R&D that could lead to U.S. production of a new nuclear weapon in phase 1, 2, or 2a, or a modified nuclear weapon in phase 6.1, 6.2, or 6.2a, or concept work occurring before phase 1 or 6.1; and (2) to request funds for each weapon activity in phase 3 or 6.3 or higher as a separate line item. Further, the legislation provided a clear definition of a key term:

> (3) The term "new nuclear weapon" means a nuclear weapon that contains a pit or canned subassembly, either of which is neither —
>
> (A) in the nuclear weapons stockpile on the date of the enactment of this Act; nor
> (B) in production as of that date.[23]

While the move from phase 2a or 6.2a to phase 3 or 6.3 typically involves a large increase in cost, requiring each weapon in phase 3 or 6.3 to have its own line item enhances the visibility of these programs and prevents them from being lumped with other programs, aiding congressional oversight and control of the process.

The explanatory language of this section of the conference report referenced the Spratt-Furse provision: "The conferees agree that nothing in this section may be construed to modify, repeal, or in any way affect the provisions of section 3136 of the National Defense Authorization Act for Fiscal Year 1994 ...."[24]

LEGISLATIVE ACTIONS IN THE FY2004 BUDGET CYCLE

The Administration, in DOD legislative proposals for FY2004, requests repeal of the Spratt-Furse provision restricting low-yield nuclear weapon R&D. Its draft language states simply, "Section 3136 of the National Defense Authorization Act for Fiscal Year 1994 (P.L. 103-160; 107 Stat. 1946) is repealed."[25] The Administration offered the following rationale:

---

23 A pit is the fissile core of a nuclear weapon. The pit and associated materials are called the primary stage. Detonating the primary provides the energy to detonate the canned subassembly, a term used to refer to the secondary stage. That stage provides most of a weapon's energy through fission and fusion.

24 U.S. Congress. Committee of Conference. *National Defense Authorization Act for Fiscal Year 2003.* 107th Congress, 2nd Session, H.Rept. 107-772, USGPO, 2002, p. 786.

25 For the proposed bill, see [http://www.defenselink.mil/dodgc/lrs/docs/March3-bill.pdf].

> Section 3136, the so-called PLYWD legislation, prohibits the Secretary of Energy from conducting any research and development which could potentially lead to the production by the United States of a new low-yield nuclear weapon, including a precision low yield warhead.
>
> This legislation has negatively affected U.S. Government efforts to support the national strategy to counter WMD and undercuts efforts that could strengthen our ability to deter, or respond to, new or emerging threats.
>
> A revitalized nuclear weapons advanced concepts effort is essential to: (1) train the next generation of nuclear weapons scientists and engineers; and (2) restore a nuclear weapons enterprise able to respond rapidly and decisively to changes in the international security environment or unforeseen technical problems in the stockpile. PLYWD has had a "chilling effect" on this effort by impeding the ability of our scientists and engineers to explore the full range of technical options. It does not simply prohibit research on new, low-yield warheads, but prohibits any activities "which could potentially lead to production by the United States" of such a warhead.
>
> It is prudent national security policy not to foreclose exploration of technical options that could strengthen our ability to deter, or respond to, new or emerging threats. In this regard, the Congressionally-mandated Nuclear Posture Review urged exploration of weapons concepts that could offer greater capabilities for precision, earth penetration (to hold at risk deeply buried and hardened bunkers), defeat of chemical and biological agents, and reduced collateral damage. The PLYWD legislation impedes this effort.

Repeal of PLYWD, however, falls far short of committing the United States to developing, producing, and deploying new, low-yield warheads. Such warhead concepts could not proceed to full-scale development, much less production and deployment, unless Congress authorizes and appropriates the substantial funds required to do this.[26]

The Senate bill, S. 1050, as reported from the Armed Services Committee, contained this repeal (Section 3131) and added that nothing in the repeal "shall be construed as authorizing the testing, acquisition, or deployment of a low-yield nuclear weapon." On May 20, the Senate considered an amendment by Senators Feinstein and Kennedy to rescind Section 3131.[27] The debate, in which more than 20 Senators participated, aroused considerable interest. Senator Feinstein argued that repealing the low-yield ban "will ... begin a new era of nuclear proliferation."[28] Senator Durbin said, "This bill is about to discard 50

---

[26] U.S. Department of Defense. "DoD Legislative Proposals: National Defense Authorization Act for Fiscal Year 2004 (Sent to Congress on March 3, 2003) —Section-by-Section Analysis." [http://www.defenselink.mil/dodgc/lrs/docs/March3-sectional.pdf]

[27] For further details on congressional action on the low-yield ban, ACI, and RNEP in 2003, see John Isaacs, "Congress Jumps off the Ban-Wagon," *Bulletin of the Atomic Scientists,* July-August 2003: 20, 21, 69.

[28] U.S. Congress. Congressional Record, May 20, 2003: S6663.

years of American foreign policy and 50 years of American nuclear policy."[29] Senator Kyl said, "The reason low-yield weapons research is being sought is because the world has changed since the time we developed these huge megaton nuclear weapons ... the United States would prefer, if it had to, to use a much smaller weapon, a low-yield weapon."[30] The amendment was tabled, 51-43.[31] When that amendment failed, Senator Jack Reed offered an amendment to modify Section 3131 by replacing "research and development" with "development engineering." Thus, the amendment would have barred development engineering that could lead to U.S. production of a new nuclear warhead. Senator Warner offered an amendment to the Reed amendment, replacing the latter with the original committee amendment and adding a section, "The Secretary of Energy may not commence the engineering development phase or any subsequent phase of a low-yield nuclear weapon unless specifically authorized by Congress."[32] The difference between the two amendments was summed up as follows:

> Mr. Reed. Essentially, the functional difference between my amendment and your second degree is, at this point, under my amendment the administration would have to come and lift the prohibition; under your amendment, they would have to come and get an authorization. I think that is the functional difference.
>
> Mr. Warner. I think the Senator is correct.[33]

On May 21, the Senate agreed to the Warner amendment, 59-38, and agreed to the Reed amendment, as amended by the Warner amendment, 96-0. Subsequently, on May 22, the Senate passed S. 1050, as amended, 98-1, and on June 4 passed H.R. 1588, the House version of the National Defense Authorization Act for FY2004, striking all after the enacting clause and inserting the text of S. 1050.

The House Armed Services Committee tackled the same subject. As reported by the committee, Section 3111 of H.R. 1588 modified the low-yield provision substantially from the language the Administration had requested,

---

[29] Ibid., p. S6669.

[30] Ibid., p. S6676.

[31] For amendment text, debate, and vote, see U.S. Congress. *Congressional Record,* May 20, 2003, p. S6663, S6663-S6690, and S6690, respectively.

[32] For text, debate, and vote on the two amendments, see U.S. Congress. *Congressional Record,* May 20, 2003: S6690-S6691; May 20, 2003: S6691-S6696 and May 21, 2003: S6789-S6792; and May 21, 2003: S6792, respectively.

[33] U.S. Congress. *Congressional Record,* May 20, 2003: S6695.

adopting by voice vote an amendment by Representative Spratt.[34] The FY1994 provision is as follows:

> (b) LIMITATION- The Secretary of Energy may not conduct, or provide for the conduct of, research and development which could lead to the production by the United States of a low-yield nuclear weapon which, as of the date of the enactment of this Act, has not entered production.

Section 3111 would revise that provision:

> (b) LIMITATION- The Secretary of Energy may not develop, produce, or provide for the development or production of a low-yield nuclear weapon which, as of November 30, 1993, has not entered production. ...
> (d) Effect on Studies and Design Work — Nothing in this section shall prohibit the Secretary of Energy from conducting, or providing for the conduct of, concept definition studies, feasibility studies, or detailed engineering design work.

The amendment apparently drew the line separating permitted from prohibited activities between phase 2a/6.2a and phase 3/6.3. The explanatory statement of the committee was somewhat clearer on this point: "The amendment would maintain the prohibition on development of new nuclear weapons with yields less than five kilotons, but would allow research on such weapons, including concept definition studies, feasibility studies, and detailed engineering design."[35] A statement of additional views by Representative Spratt and 24 other Democrats, in the committee's report on the bill, elaborated: "the amendment permits research of such [sub-5-kt] weapons, it prohibits development engineering (referred to as Phase 6.3 activities by the Department of Energy) and later stages of development."[36]

The amendment is more permissive than the original provision but less so than the one the Administration had proposed, apparently permitting work through phase 2a/6.2a but barring subsequent development. It also narrows the potentially broad applicability of the phrase "which could lead to the production." That phrase raises the prospect that a paper study, a computer model, or a metallurgical experiment might be illegal if it could somehow be construed as leading to production. With the new language, that possibility would no longer be enough to block such work; rather, the test would be

---

[34] David Morris, "Markup Reports: Panel Extends Pentagon Power over Personnel, Wildlife," *CongressDaily,* May 14, 2003: 2.

[35] House Armed Services Committee, National Defense Authorization Act for Fiscal Year 2004. p. 434.

[36] Ibid., p. 519.

whether the development or production is of an actual low-yield weapon. The rule for H.R. 1588 did not provide for amendments to this provision. The House passed the bill, as amended, on May 22, 361-68.

CONFERENCE ISSUES AND OUTCOMES

Several aspects of the low-yield ban were potentially at issue in conference. Both bills sought to define which activities were permitted and which were barred in order to remove an ambiguity of the original legislation: the issue of R&D that "could lead to the production by the United States of a new low-yield nuclear weapon, including a precision low-yield warhead." They approached the problem differently, though. The Senate would have rescinded the low-yield R&D provision while barring "engineering development" or subsequent phases of a low-yield weapon without congressional approval; the House would have retained the lowyield ban while modifying it substantially. The Senate bill, as amended, clearly drew the line separating permitted activity and activity requiring explicit congressional authorization between phase 2a (or 6.2a) and phase 3 (or 6.3). The House language appeared to do the same thing, but it was not completely clear that "develop" and "detailed engineering design work" referred to that same line. It was unclear whether there were a difference between the Senate's requirement for "explicit authorization" of engineering development or beyond and the requirement in Section 3143 of P.L. 107-314 for DOE to include a line-item request for each new or modified nuclear weapon in phase 3 or 6.3 or beyond. A senior NNSA staff member held that while committee or conference report language would typically comment on weapons in phase 3 or 6.3 or beyond, NNSA would view authorization and appropriations of funds for such weapons, typically hundreds of millions of dollars, as constituting the "explicit authorization" that the Senate's bill required, but that in any event it would be up to Congress to define what form of authorization sufficed.[37]

The House agreed to the conference report,[38] 362-40, on November 7, 2003. The Senate agreed to the conference report, 95-3, on November 12. The President signed the measure into law (P.L. 108-136) on November 24.

The conference bill, section 3116, adopted the Senate provision. Section 3116 repealed the Spratt-Furse provision, stated that the repeal shall not "be construed as authorizing the testing, acquisition, or deployment of a low-yield

---

[37] Telephone conversation, June 10, 2003.

[38] U.S. Congress. Committee of Conference. *National Defense Authorization Act for Fiscal Year 2004.* H.Rept. 108-354, 108th Congress, 1st Session, USGPO, 2003.

nuclear weapon," and barred the Secretary of Energy from beginning engineering development, or subsequent phases, of work on a low-yield weapon without specific congressional authorization. The bill further required a report by the Secretaries of State, Defense, and Energy assessing whether or not the repeal of Spratt-Furse "will affect the ability of the United States to achieve its nonproliferation objectives and whether or not any changes in programs and activities would be required to achieve those objectives."

The legislation thus clearly delineates what R&D on low-yield nuclear weapons is prohibited, eliminating a point of confusion in the original provision, and sets the requirement (congressional approval) for overriding that prohibition. At the same time, the requirement for an assessment of "any changes in programs and activities" needed to achieve U.S. nonproliferation objectives is unclear both in terms of what those objectives are and how broadly the phrase "any changes" is to be interpreted. Such changes could range from developing new low-yield nuclear weapons, to training more specialists in Arabic, to strengthening the International Atomic Energy Agency. And what definition of nonproliferation objectives is the report to use?[39]

DEVELOPMENTS IN THE FY2005 BUDGET CYCLE

It appears that low-yield R&D will be a much less salient issue in the FY2005 cycle than in the previous one. Lifting the Spratt-Furse provision was a policy issue that was resolved last year, and not a budget issue. Some critics expressed concern that NNSA would use the lifting of the provision to develop new weapons. In an interview of December 2003, however, NNSA Administrator Linton Brooks stated, "there is no list of low-yield weapons we're thirsting to develop – that's a misconception."[40] An NNSA manager with responsibilities in this area elaborated in March 2004:

> We have no requirements from DoD for low yield weapon research. ... We definitely do not have funds identified specifically for that area of research. ... We may get other requests between now and the start of FY2005. One or more could be for concepts that would delve into the low yield area. So I can't and

[39] The *National Strategy to Combat Weapons of Mass Destruction* (U.S. White House. December 2002), for example, lists such nonproliferation tools as active nonproliferation diplomacy, multilateral regimes, nonproliferation and threat reduction cooperation, and controls on nuclear materials (p. 4-5), but does not address goals except in the most general terms.

[40] "The Bush Administration's Views on the Future of Nuclear Weapons: An Interview with NNSA Administrator Linton Brooks," *Arms Control Today,* January/February 2004: 3.

won't claim that none will be done in FY2005, just that as of today, we aren't planning any.[41]

ISSUES FOR CONGRESS

Would Lifting the Ban Lead to Acquisition and Use of Nuclear Weapons?

Supporters of the ban argue that new weapons would make acquisition and use of nuclear weapons more likely. Any new weapon would broaden the President's range of options. A President might find options made available by current nuclear weapons to be unacceptable, but a new weapon might offer a tradeoff between costs and benefits that tilts in favor of nuclear use. Critics further believe that the Administration is seeking to lift the ban with the intention of building lowyield weapons. Senator Feinstein asked Secretary of Defense Donald Rumsfeld about it, "and he said it's just a study .... They did the same thing with the nuclear posture review: Oh, it's just an intellectual exercise. I don't believe either of those, not one whit, and I think there's a very clear march on to develop these weapons."[42] Critics further argue that even research on low-yield weapons would lead down a slippery slope to testing, production, deployment, and use. Senator Feinstein said, "The repeal of Spratt-Furse opens the door for America to begin to develop nuclear weapons again."[43] Senator Biden said that many supporters of the ban are concerned because "we believe very much that if one of these weapons ... is developed, it will ultimately be fielded," and that low-yield weapons send a "dangerous signal ... to other countries, whether intentional or not, that we intend to fight a nuclear war."[44] Senator Kennedy said, "a mini-nuke is still a nuke. ... If we build it, we will use it. ... it is a one-way street that can lead only to nuclear war."[45] Critics prefer to stop the momentum in this direction at the earliest opportunity.

Those who would lift the ban see "lowering the nuclear threshold" and a "slippery slope to nuclear use" as a misreading of U.S. policy and practice. Congress will decide if a weapon is to move to phase 3/6.3 through the authorization and appropriations process; approval would be far from automatic.

---

[41] Information provided by NNSA staff, March 4, 2004.

[42] Senators Edward Kennedy and Dianne Feinstein, press conference, "A New Proliferation of Nuclear Weapons," September 16, 2003, transcript by Federal News Service.

[43] U.S. Congress. *Congressional Record,* May 20, 2003: S6665.

[44] Ibid., p. S6673 and S6682.

[45] Ibid., p. S6687.

The provision repealing Spratt-Furse in the Senate Armed Services Committee's bill (S. 1050, Sec. 3131) stated explicitly, "Nothing in the repeal ... shall be construed as authorizing the testing, acquisition, or deployment of low-yield nuclear weapons." Senator Kyl stated, "this is not an authorization. All we are doing is removing a self-imposed restriction on thinking about this, on doing research."[46]

Further, they note, use of nuclear weapons has always been a presidential decision, and one that Presidents treat as arguably the most weighty decision that they can make. Having more weapons, or low-yield weapons, does not mean that the United States will use them. As Senator Warner observed,

> the threshold for using nuclear weapons remains very high indeed. ... the United States had a large number of low-yield nuclear weapons in our inventory during the '50s, '60s, and '70s which have now been removed from the inventory. During each of these decades there were significant national security challenges to the United States. None of those challenges came close to reaching the nuclear threshold ...[47]

Would Low-Yield Weapons Make Nuclear Proliferation More Likely?

Those who would retain the ban assert that resuming low-yield R&D could spur nuclear proliferation. If the world's only superpower requires for its security new types of nuclear weapons in addition to the ones it already has, then this implies that other nations need nuclear weapons for their security as well. Building weapons that might be used against rogue states will not deter these states from building nuclear weapons, it is argued, but will instead lead them to develop such weapons of their own to deter U.S. attack. They maintain that although nonproliferation efforts have not been 100 percent successful, they have served as an important restraint.

Resuming low-yield R&D could lead to nuclear testing, critics argue, likely leading to resumed nuclear testing by others. The head of the nuclear directorate of Russia's defense ministry reportedly stated in January 2002, "if any of the five countries officially possessing nuclear weapons starts testing nuclear munitions again, and we consider our nuclear stockpile to be in a critical state, we too will carry out nuclear tests."[48] Beyond that, the NPT is the cornerstone of

---

[46] Ibid., p. S6677.

[47] Ibid., p. S6687.

[48] Letter from Representative Edward Markey and 39 other Representatives to Representatives Bob Stump and Ike Skelton on Robust Nuclear Earth Penetrator and

the nuclear nonproliferation regime, which incorporates other agreements, treaties, groups, and arrangements as well.[49] The NPT faced a key decision in 1995: states party had to decide whether or not to extend the treaty indefinitely. They decided to do so, but this extension was linked to a set of nonproliferation objectives, one of which was completing negotiations on a CTBT no later than 1996. Accordingly, as Sidney Drell and others argue, "A decision to resume [nuclear] testing to build low-yield nuclear weapons could deal the [nonproliferation] regime a fatal blow ..."[50]

Those opposed to the ban reject the argument that studying low-yield weapons would lead to nuclear proliferation. No other nation has such a prohibition, so lifting it would simply move the United States to a position of parity in that regard, they assert. Nations will act in their own interests on decisions involving their security or ambitions. The United States signed the CTBT in 1996, yet India and Pakistan tested nuclear weapons in 1998. These critics question the NPT's utility. North Korea and (apparently) Iran made substantial progress on their nuclear programs even while being party to that treaty. Nuclear programs in all four nations were underway decades ago, long before the U.S. ban on low-yield R&D. The ban's opponents argue that U.S. actions — especially on something as remote as preliminary studies of possible weapons — will not make other nations more likely to develop WMD. To the contrary, they believe, U.S. ability to respond to WMD proliferation efforts by rogue states may dissuade them from undertaking such actions in the first place. As Senator Domenici said, "to permit [weapons scientists] to work in this area [lowyield weapons] is part of the deterrent."[51]

The ban's opponents reject the contention that R&D on new weapons will lead to nuclear testing. As C. Paul Robinson, Director of Sandia National Laboratories, said, "I can categorically state that no one is proposing returning to nuclear testing."[52] The United States has decades of experience with low-yield nuclear weapons, including those with altered radiation outputs, such as the neutron bomb of the 1970s. Use of existing, tested designs, perhaps with modifications not requiring testing, would provide high confidence. A better understanding of the effects of low-yield weapons is crucial for understanding

nuclear test readiness, May 1, 2002.

[49] For further information on the nuclear nonproliferation regime, see CRS Report RL31559, *Proliferation Control Regimes: Background and Status*, by Sharon Squassoni, Steven Bowman, and Carl Behrens.

[50] Sidney Drell, James Goodby, Raymond Jeanloz, and Robert Peurifoy, "A Strategic Choice: New Bunker Busters Versus Nonproliferation," *Arms Control Today,* March 2003: 8.

[51] U.S. Congress. *Congressional Record,* May 20, 2003: S6674.

[52] James Kitfield, "The Pros and Cons of New Nuclear Weapons," *National Journal,* August 9, 2003: 2567.

their utility, yet key variables are how a blast wave would interact with structures; computer models and engineering studies would arguably provide the needed data.

Would Low-Yield Weapons Offer Military Value?

Some supporting the ban argue there is little meaningful use for low-yield weapons. They note a statement by Linton Brooks, at the time Acting Administrator of NNSA: "we have no requirement to actually develop any new weapons at this time."[53] As the Afghanistan and Iraq Wars showed, it is often difficult to know what to target. Al Qaeda presents few if any targets that would be suitable for nuclear weapons. Another lesson of these wars is that if targets can be located, they can often be defeated with conventional weapons. Deep burial, large tunnel complexes with multiple barriers, deception, etc., might make targets in rogue states difficult to destroy with nuclear weapons even if they can be located. Many potential targets are in cities, but detonating a low-yield nuclear weapon in or even near a city could cause much collateral damage. By one estimate, a 5-kiloton weapon detonated near and upwind from Damascus, Syria, at a depth of 30 feet would cause 230,000 fatalities and another 280,000 casualties within two years.[54] Use of a low-yield earth penetrator against the bunkers thought to house Saddam in Baghdad, a city of nearly 5 million people, could have caused casualties on a similar scale.

The ban's opponents claim that low-yield weapons may offer important military capabilities, such as for destroying biological munitions (discussed under ACI, below) or attacking underground bunkers. There is military interest in low-yield weapons. Admiral James Ellis, Commander, U.S. Strategic Command, wrote that "US Strategic Command is interested in conducting

---

[53] "Statement of Linton F. Brooks, Acting Under Secretary of Energy and Administrator for National Security, National Nuclear Security Administration, U. S. Department of Energy Before the Subcommittee on Strategic Forces, Committee on Armed Services, U. S. Senate, April 8, 2003," p. 6.: [http://www.senate.gov/~armed_services/statemnt/2003/April/Brooks.pdf]

[54] U.S. Congress. *Congressional Record,* May 20, 2003: S6666. Senator Kennedy cited this estimate. Matthew McKinzie, staff scientist on the nuclear program of the Natural Resources Defense Council, performed the analysis in April 2003 using the HPAC ("Hazard Prediction Assessment Capability") computer code produced by SAIC, Inc., for the Defense Threat Reduction Agency. Telephone conversation with Matthew McKinzie, September 23, 2003. For further information on the code and calculations using it, see McKinzie's statement in "New Nuclear Weapons vs. Nonproliferation: The Choice Before Congress," press briefing held by the Arms Control Association, Washington, DC, April 29, 2003: [http://www.armscontrol.org/events/newnuclearweapons_apr03.asp]

rigorous studies of all new technologies, and examining the merits of precision, increased penetration, and reduced yields for our nuclear weapons."[55] At present, in the view of those who would lift the ban on low-yield R&D, the ban makes it impossible to assess the value of the capabilities sub-5-kt weapons might bring to the stockpile, or how they might be designed for specific missions. The ban, they assert, imposes broader constraints on nuclear weapons designers as they seek to stay well away from any activities that could be construed as violating the Spratt-Furse provision. Senator Sessions concluded, "it would be irrational to prohibit research that could inform future decisions as to whether such weapons would enhance the national security of our country."[56] Senator Domenici noted that the ban increases the leadtime to develop a low-yield weapon should the need arise.[57]

---

[55] Letter from Admiral Ellis to Senator John Warner, (no date), in U.S. Congress. *Congressional Record,* May 20, 2003: S6680.

[56] U.S. Congress. *Congressional Record,* May 20, 2003: S6667.

[57] Ibid., p. S6675.

Chapter 4

ADVANCED CONCEPTS INITIATIVE

DESCRIPTION

For FY2004, the Administration requested $21.0 million for the Advanced Concepts Initiative (ACI). Of this amount, $15.0 million was for the Robust Nuclear Earth Penetrator (RNEP) weapon, discussed in the next section, and $6.0 million ($2.0 million for each of the three weapons labs) was for other advanced weapon concepts, discussed here. This latter program involves studies through phase 2a/6.2a by small groups of weapons personnel at Los Alamos, Livermore, and Sandia, as well as liaison with DOD commands. A Los Alamos publication stated:

> The ACI program will conduct new weapon studies and will explore concepts for new warhead designs and modifications to meet DoD needs that are not met by the current stockpile. The ACI is a program for developing and exercising capability and for applying that capability to examine options.[1]

In its FY2004 legislative proposals, the Administration argued for ACI:

> A revitalized nuclear weapons advanced concepts effort is essential to: (1) train the next generation of nuclear weapons scientists and engineers; and (2) restore a nuclear weapons enterprise able to respond rapidly and decisively to changes in the international security environment or unforeseen technical problems in the stockpile.[2]

---

[1] John Immele, "Los Alamos and the New Triad," *Nuclear Weapons Journal,* May/June 2003: 31.

[2] U.S. Department of Defense. *DoD Legislative Proposals. National Defense Authorization Act for Fiscal Year 2004.* Sent to Congress March 3, 2003. Section-by-Section Analysis, p. 2. [http://www.defenselink.mil/dodgc/lrs/docs/March3-sectional.pdf]

The FY2004 NDAA and EWDAA both provided $6.0 million, as requested, but conferees on the latter made $4.0 million available for authorization only after delivery of a DOE-DOD report on the nuclear stockpile and a 90-day congressional review period. As of early March 2004, the report had not been delivered, and NNSA could offer no indication as to when that might occur. As a result, ACI may be limited to a budget of $2 million for much if not all of FY2004. The FY2005 request for ACI is $9.0 million.

HISTORY THROUGH THE FY2003 BUDGET CYCLE

During much of the Cold War, the three weapons laboratories had teams of scientists and engineers studying advanced weapons concepts. Studies ranged from modifications of existing weapons, to improvements for next-generation weapons, to exploration of new weapons technologies and weapons for new missions. Tools used included laboratory experiments, computer modeling, data from past and ongoing nuclear tests, and data from experiments the teams added to nuclear tests. These teams were small, perhaps a half-dozen full-time professional staff, but during the Cold War they drew on the large advanced and exploratory R&D program that was a major part of the core nuclear weapons programs at the laboratories. Team members, new staff as well as experienced designers, would typically work on advanced concepts for a year or two. Team members would also interact with DOE headquarters, DOD elements, contractors, and others.

With the end of the Cold War and the end of nuclear weapons development, the laboratories wound down their advanced concepts programs in the middle 1990s. Since then, there have been statements emphasizing the value of the type of work that advanced concepts programs performed, and subsequent calls for an ACI to reestablish advanced concepts programs at the labs. ACI would recreate the small teams, but would not reorient the bulk of the nuclear weapons program, which is focused on sustaining current weapons through the Stockpile Stewardship Program. The Chiles report of 1999 stated, "DOE should also encourage the laboratories to continue their decades old practice of exploratory development programs since these programs have allowed experienced engineers and scientists to maintain their systems engineering skills and train new employees."[3] The Foster panel's FY1999 report stated: "The nuclear weapons complex should work on a range of design and development tasks that

[3] U.S. Commission on Maintaining United States Nuclear Weapons Expertise. *Report to Congress and the Secretary of Energy.* March 1, 1999, p. 32. [http://www.fas.org/nuke/guide/usa/doctrine/doe/chilesrpt.pdf]

exercise and sustain the capability to produce new weapon designs. This provides both a broader set of technical options to meet future needs, and a program for training new generations of stewards."[4]

The purportedly leaked version of the Nuclear Posture Review stated:

> There are several nuclear weapon options that might provide important advantages for enhancing the nation's deterrence posture: possible modifications to existing weapons to provide additional yield flexibility in the stockpile; improved earth penetrating weapons (EPWs) to counter the increased use by potential adversaries of hardened and deeply buried facilities; and warheads that reduce collateral damage.
>
> To further assess these and other nuclear weapons options in connection with meeting new or emerging military requirements, the NNSA will reestablish advanced warhead concepts teams at each of the national laboratories and at headquarters in Washington. This will provide unique opportunities to train our next generation of weapon designers and engineers. DoD and NNSA will also jointly review potential programs to provide nuclear capabilities, and identify opportunities for further study, including assessments of whether nuclear testing would be required to field such warheads.[5]

The Administration now proposes to re-create these teams.

In July 2003, NNSA provided details on the FY2004 studies that the non-RNEP part of ACI would conduct.[6] Perhaps $200,000 would go for participation by the weapons labs in an Air Force concept study of enhancing and modernizing a current warhead, such as examining if any additional military characteristics are useful and feasible. The balance would go for other laboratory studies. Specifically, Los Alamos would spend its funds on enhancing computer modeling and simulation capability and for mechanical testing. Sandia would continue a study of "mission end-to-end command and control," which would enable military personnel to know where a warhead was at all times. This would be applicable to conventional as well as nuclear weapons. For example, if a laser-guided bomb lost track of its laser signal because of fog or smoke, end-to-end command and control would indicate its location. Livermore would undertake an early study on "design to effect," i.e., designing nuclear weapons to obtain desired output characteristics. These projects may or may not support research directly on low-yield nuclear weapons.

---

4 John Foster, Chairman, et al., FY 1999 Report of the Panel to Assess the Reliability, Safety, and Security of the United States Nuclear Stockpile, November 8, 1999, p. 12.

5 Nuclear Posture Review [Excerpts], p. 34-35.

6 Information provided by a DOE official, July 23, 2003.

LEGISLATIVE ACTIONS IN THE FY2004 BUDGET CYCLE

For FY2004, the Senate Armed Services Committee recommended authorizing the full $21.0 million requested for Advanced Concepts.[7] The House Armed Services Committee stated that ACI had several important purposes, such as exercising the weapons design process, training the next generation of nuclear weapons scientists and engineers, and understanding what adversaries might do in the area of nuclear weapon design. Accordingly, the committee expressed its "[belief] that NNSA should consider more significant future investment in these [Advanced Concepts] activities" and "recommends that NNSA proceed with its advanced concepts initiative forthwith."[8] No amendments were offered on non- RNEP ACI in the Senate. The House rejected an amendment by Representative Tauscher to transfer all funds from ACI (including funds for RNEP) to conventional programs to defeat hardened and deeply buried targets, as discussed under RNEP, below. As a result, non-RNEP ACI was not at issue in the conference.

In its report on H.R. 2754, the FY2004 Energy and Water Development Appropriations Bill, the House Appropriations Committee delivered a scathing criticism of the nuclear weapons program. The committee found that a flawed budget process raised questions about the legitimacy of nuclear weapon requests:

> Unfortunately, the country possesses neither a modern stockpile nor a modern nuclear weapons complex. Instead, both are largely carryovers from the Cold War era. After careful consideration, the Committee has concluded that much of the current situation results from a flawed budget process. ... the weapons activities portion of the NNSA budget is effectively insulated from any such [intra-agency] tradeoffs — DoD sets requirements that another agency has to fund, and DOE treats the weapons activities budget as untouchable because DoD set the requirements.
>
> There needs to be a serious debate about whether the approximately $6 billion spent annually on DOE's nuclear weapons complex is a sound national security investment. Until that debate occurs and the DOE weapons budget request is subject to meaningful budget trade-offs, this Committee will not assume that all of the proposed nuclear weapons requests are legitimate requirements.[9]

---

[7] U.S. Congress. Senate. Committee on Armed Services. *National Defense Authorization Act for Fiscal Year 2004,* S.Rept. 108-106, p. 442.

[8] U.S. Congress. House. Committee on Armed Services. *National Defense Authorization Act for Fiscal Year 2004,* H.Rept. 108-106, p. 429-430.

[9] U.S. Congress. House. Committee on Appropriations. *Energy and Water Development Appropriations Bill, 2004.* H.Rept. 108-212, 108th Congress, 1st Session, USGPO, 2003, p. 139-140.

This disconnect, the committee stated, led to new weapon-related initiatives, such as ACI, RNEP, and enhanced test readiness, that overstretched NNSA's management abilities.

> It appears to the Committee the Department [of Energy] is proposing to rebuild, restart, and redo and otherwise exercise every capability that was used over the past forty years of the Cold War and at the same time prepare for a future with an expanded mission for nuclear weapons. Nothing in the past performance of the NNSA convinces this Committee that the successful implementation of Stockpile Stewardship program is a foregone conclusion, which makes the pursuit of a broad range of new initiatives premature. Until the NNSA has demonstrated to the Congress that it can successfully meet its primary mission of maintaining the safety, security, and viability of the existing stockpile by executing the Stockpile Life Extension Program and Science-based Stewardship activities on time and within budget, this Committee will not support redirecting the management resources and attention to a series of new initiatives.[10]

For these reasons, it recommended eliminating the $6.0 million requested for ACI. The House passed H.R. 2754 on May 18, 2003, 377-26; no amendments were offered on ACI, RNEP, test readiness, or any other Weapons Activities provisions. (Weapons Activities is the part of NNSA's budget that funds stockpile stewardship.)

The Senate Appropriations Committee recommended providing the full amount requested for ACI.[11] The Senate considered several amendments on ACI on September 16. It tabled, 53-41, an amendment by Senator Feinstein to eliminate the full $21.0 million for RNEP and other ACI, and for other purposes.[12] It adopted on voice vote an amendment by Senator Jack Reed that barred use of funds provided by H.R. 2754 for phase 3 or 6.3 or beyond for advanced nuclear weapon concepts including RNEP.[13] Senator Reed stated that his amendment "would assure that the appropriations bill is consistent with the language [on RNEP by Senator Nelson of Florida] that is included in the fiscal year 2004 Defense authorization bill."[14] The Senate agreed to various amendments *en bloc,* including one by Senator Bingaman that barred spending

---

[10] Ibid., p. 145.

[11] U.S. Congress. Senate. Committee on Appropriations. *Energy and Water Development Appropriation Bill, 2004.* S.Rept. 108-105, 108th Congress, 1st Session, USGPO, 2003, p. 107.

[12] For amendment text, debate, and vote, see U.S. Congress. *Congressional Record,* September 15, 2003: S11436; September 15, 2003: S11436-S11451 and September 16, 2003: S11531-S11533; and September 16, 2003: S11533, respectively.

[13] For amendment text, debate, and vote, see U.S. Congress. Congressional Record,September 16, 2003: S11534, S11534-S11538, and S11538, respectively.

[14] U.S. Congress. *Congressional Record,* September 16, 2003: S11535.

funds provided by the bill on additional and exploratory studies (a category that excludes RNEP) under ACI until 30 days after NNSA gives Congress a detailed report on activities planned in that category for FY2004.[15] The Senate passed the bill, H.R. 2754, on September 16, 92-0.

CONFERENCE ISSUES AND OUTCOMES

Non-RNEP ACI was not expected to be at issue in the defense authorization conference because each House provided the same amount for it. Senate conferees could, however, have considered whether to endorse the House language regarding the program's future budget and proceeding, modify it, or be silent on it. Because of congressional interest in nuclear weapon programs, conferees could also have considered directing DOE to specify in detail the projects that future-year ACI requests will support.

Non-RNEP ACI was at issue in the energy and water conference. A point of possible contention was whether conferees could arrive at a compromise somewhere between the House and Senate figures, or if the funding request was so small relative to other NNSA projects that a cut of, say, 50 percent would prevent one or two of the three labs from performing its planned study. Also at issue were the two amendments the Senate passed.

An issue that implicitly faced energy and water conferees was the gulf between the House Appropriations Committee report, which argued that "pursuit of a broad range of new [nuclear] initiatives [was] premature," and the House Armed Services Committee report, which urged NNSA to "consider more significant future investment" in ACI (including RNEP).[16] Further, the authorization bill as passed by the House included the $6.0 million requested, while the appropriations bill as passed by the House eliminated those funds.

The section on the low-yield R&D ban, above, gave details on final passage of the NDAA. Regarding the EWDAA, the House agreed to the conference report, 387-36, on November 18, and the Senate agreed to it on the same day by unanimous consent. The President signed the measure into law (P.L. 108-137) on December 1.

---

[15] For amendment text and *en bloc* agreement, see U.S. Congress. *Congressional Record,* September 16, 2003: S11541 and S11539, respectively.

[16] U.S. Congress. House. Committee on Appropriations. *Energy and Water Development Appropriations Bill, 2004*, H.Rept. 108-212, p. 145; and U.S. Congress. House. Committee on Armed Services. *National Defense Authorization Act for Fiscal Year 2004,* S.Rept. 108-106, p. 429.

The NDAA provided the amount requested for non-RNEP ACI, $6.0 million. It was not at issue in conference, and the conference report did not comment on it.

The EWDAA provided $6.0 million for Advanced Concepts (excluding RNEP), but the conferees made $4.0 million available for obligation only after the Secretaries of Energy and Defense deliver to Congress a revised Nuclear Weapons Stockpile plan detailing a plan and schedule for achieving the President's proposed adjustments to the strategic weapons stockpile (including a reduction in operationally deployed weapons to 1,700-2,200 by 2012), and 90 days elapse for review by the Armed Services and Appropriations Committees.[17] The DOE-DOD report was due with the submission of the FY2005 budget request in early February 2004, but the two departments had not submitted it a month later.[18] This provision therefore limits ACI funds to $2.0 million until the report is submitted. Further, conferees included as section 313 the amendment by Senator Bingaman barring spending on ACI until 30 days after NNSA submits a report detailing certain planned ACI activities. NNSA submitted the report on March 3, 2004, so the activities are suspended until April 2.[19]

While the EWDAA provided the Senate amount for ACI and not the House amount, the conferees' language requiring a DOE-DOD report on the link between the stockpile plan and the President's planned nuclear force structure addressed a central concern expressed in the House Appropriations Committee's report:

> While the conventional forces in the Defense Department go through a 21st-Century transformation to meet the challenges of a new era, the NNSA is forced, through inertia and indecision, to maintain all contingencies regardless of how unlikely the threat. The Department of Defense needs to determine the composition of the stockpile required to support the President's announced stockpile reductions, and then coordinate with DOE to establish the nuclear weapons complex requirements based on deliberate, timely, well-justified decisions supported by Congress.[20]

Regarding the amendment by Senator Reed on RNEP, the conference report contained the following language: "The conferees remind the Administration that none of the funds provided may be used for activities at the engineering

---

17 U.S. Congress. Committee of Conference. Making Appropriations for Energy and Water Development for the Fiscal Year Ending September 30, 2004, and for Other Purposes. H.Rept. 108-357, 108th Congress, 1st Session, USGPO, 2003, p. 154, 156.

18 Information provided by NNSA staff, March 4, 2004.

19 Information provided by NNSA staff, March 4, 2004.

20 House Appropriations Committee, *Energy and Water Development Appropriations Bill, 2004,* H.Rept. 108-212, p. 142.

development phases, phase 3 or 6.3, or beyond, in support of advanced nuclear weapons concepts, including the Robust Nuclear Earth Penetrator."[21]

THE FY2005 BUDGET CYCLE

The FY2005 request is $9.0 million. ACI activities include

> developing advanced concepts which could be applied to the stockpile of the future, code development for system-specific nuclear effects, phenomenology, and exercise of design skills; conducting pre-conceptual, conceptual, feasibility, design and costing studies of options.[22]

NNSA Administrator Brooks said that ideas for ACI include the ability to destroy biological agents, improved safety and security, and more capable or robust weapon designs that might be developed without testing. He felt that the effort probably "will be focused on safety, security flexibility, greater margins than on fundamental new capabilities."[23]

NNSA's FY2005 budget request document includes, for the first time, budgets for out years (in this case FY2006-FY2009) as well as for the current year. The requests projected for ACI are: FY2006: $14.4 million; FY2007, $14.9 million; FY2008, $14.6 million; and FY2009, $29.5 million.[24]

ISSUES FOR CONGRESS

The Administration, DOD, NNSA, the nuclear weapons laboratories, the Foster panel, and others see many benefits for ACI. Many are fine-grained in their detail. The labs in particular have a detailed sense of what they would undertake through ACI; many of the arguments for ACI presented here are based on discussions with weapons lab personnel. Whether because of the relatively small amount of money involved, uncertainty over which projects ACI

---

[21] Committee of Conference, Making Appropriations for Energy and Water Development for the Fiscal Year Ending September 30, 2004 ..., p. 156.

[22] U.S. Department of Energy. Office of Management, Budget and Evaluation/CFO. *FY 2005 Congressional Budget Request,* vol. 1, National Nuclear Security Administration. DOE/ME-0032, February 2004, p. 76.

[23] "The Bush Administration's Views on the Future of Nuclear Weapons: An Interview with NNSA Administrator Linton Brooks," *Arms Control Today,* January/February 2004: 3.

[24] U.S. Department of Energy. Office of Management, Budget and Evaluation/CFO. *FY 2005 Congressional Budget Request,* vol. 1, National Nuclear Security Administration. DOE/ME-0032, February 2004, p. 63.

would undertake, or the attention drawn to RNEP, test readiness, and the ban on low-yield R&D, opponents of those three issues made few comments on ACI as a whole. Consequently, due to the paucity of public discourse on ACI, some counterarguments come from critics of the program, but others are presented as they might be made by the critics.

Does Deterrence Require New Types of Nuclear Weapons?

A key aspect of the debate over ACI — and over low-yield weapons and RNEP as well —is the relationship between deterrence and new weapons. George Miller of Livermore states:

> Deterrence is a dynamic concept — it has to continually evolve and be able to respond to changes that our adversaries will make in order to be relevant and effective. That's what happened in the Cold War. The real question that is being debated is whether nuclear weapons will be allowed to develop in response to the changing world situation. If they are to evolve, modifications to the Cold War arsenal will be required if for no other reason than to limit collateral damage. If they are not to evolve, the nuclear arsenal will become irrelevant and will die away. Standing still is not stable.[25]

Critics assert that the only role of nuclear weapons is to deter other nations from using nuclear weapons against the United States, or perhaps against its friends and allies. For this purpose, existing weapons are more than sufficient and their use is fully credible. In this view, there would be no need for new weapons.

With the threat of a U.S.-Russian nuclear war very low, the Administration is seeking something from nuclear weapons beyond deterrence. A goal set forth in the Quadrennial Defense Review is that "[w]ell targeted strategy and policy can ...dissuade other countries from initiating future military competitions."[26] The Nuclear Posture Review, amplifying on this point, stated that a "[d]iverse portfolio of capabilities denies payoff from competition."[27] The Administration further argues that preemption may be needed to counter threats of WMD, and that the United States may use "all of our options" in *responding* to an attack using WMD. While these two statements do *not* state that the United States would use nuclear weapons preemptively, neither do they rule out such use.

---

[25] Personal communication, September 5, 2003.

[26] U.S. Department of Defense. *Quadrennial Defense Review Report.* September 30, 2001, p. 12. [http://www.defenselink.mil/pubs/qdr2001.pdf]

[27] U.S. Department of Defense. *Findings of the Nuclear Posture Review,* January 9, 2002, slide, "QDR: Defense Policy Goals."

Another goal of the current Administration involves having potential adversaries take certain actions, such as having Iran open its nuclear facilities to inspection, having North Korea verifiably halt all work related to nuclear weapons, or (until recently) having Iraq cooperate fully with inspectors. Ideally, these goals would be achieved through diplomatic, economic, or similar means, but if those means do not suffice, preemptive force might be threatened or, as in Iraq, used. Coercion — the threat to use force against states if they *do not* take certain actions — goes beyond deterrence, which threatens to use force against states if they *do* take certain actions. Coercion works earlier in the threat chain, with the intention of eliminating (through diplomacy, threat, or force) activities that could create threats rather than forestalling the use of existing capabilities, one argument being that it is hard to know what is needed to deter a rogue state.

A shift from deterrence to something more, whatever form that may take, could lead to a change in weapons desired to implement the new policy. Conventional or nuclear forces credible for deterrence might not be credible under the new approach. Would any regime believe a U.S. threat to use nuclear weapons? New nuclear weapons — especially lower-yield weapons tailored to destroy specific targets —render such threats more credible, supporters argue.

ACI's supporters believe that the ability to respond to potential threats, as ACI is intended to do, would promote U.S. security. John Gordon, former NNSA Administrator, said, "a demonstrated ability to design, develop, and produce new warheads, including small builds of special purpose weapons, could be an important element in our overall deterrent posture. Such capabilities could act to convince an adversary that it could not expect to negate U.S. nuclear forces, for example, by seeking to house vital command and control functions in hard, deeply-buried installations."[28] U.S. nuclear weapons able to hold at risk such targets could be used not only for deterrence but also for coercion and preemption.

In the view of critics, deterrence and something more come from threat of regime change, and that threat comes from conventional, forces, not nuclear forces. New weapons, in this assessment, might not enhance deterrence even if tailored for specific missions, as ACI might do, because rogue states would see use of new weapons as no more credible than use of current weapons. Critics cite a 1992 letter by the first Bush Administration setting forth a presidential decision "to modify U.S. nuclear testing policy immediately, to impose limits on the number, purpose and yield of our tests. The purpose of all U.S. nuclear tests of our weapons will henceforth be for the safety and reliability of our deterrent

[28] Letter from John Gordon, Administrator, National Nuclear Security Administration, to the Honorable John W. Warner, Chairman, Committee on Armed Services, United States Senate, May 17, 2001, p. 2.

forces."[29] Omitted from this statement was weapons development as a purpose for testing. As developing new weapons typically required testing, critics interpret the statement as indicating that there was no need for new nuclear weapons. If existing weapons were sufficient for U.S. security in 1992, they assert, these same weapons should suffice now. At that time, U.S. nuclear weapons addressed as lesser included cases targets that are of current concern, such as deeply buried targets and biological weapon storage sites. Additionally, critics oppose the Administration's policy of preemption and see concurrent requests for buttressing U.S. nuclear capabilities as a dangerous part of it.

Do ACI Programs Offer Significant Military Value?

There have been several suggestions for new types of nuclear weapons in addition to earth penetrator weapons (EPWs). General Richard Myers, Chairman, Joint Chiefs of Staff, stated, "In terms of anthrax, it's said that gamma rays can, you know, destroy the anthrax spores, which is something we need to look at."[30] Kathleen Bailey, former Assistant Director for Nonproliferation, Arms Control and Disarmament Agency and Robert Barker, former Assistant to the Secretary of Defense for Atomic Energy, suggest several new warheads: "[w]arheads with suppressed electromagnetic pulse for more effective missile defense; [r]educed residual radiation warheads for low collateral damage; and [r]obust warheads for longer shelf-life."[31] Another possibility would be warheads for intercepting missiles armed with chemical or biological agents.[32] If these ideas are of mutual interest to DOE and DOD, projects of this sort would appear to fall within the purview of ACI.

Critics question the value or feasibility of such weapons. Peter Zimmerman, a physicist, consultant to the Senate Foreign Relations Committee, and formerly

---

29 Letter from [National Security Advisor] Brent Scowcroft, [Secretary of Defense] Dick Cheney, and [Secretary of Energy] James Watkins to Honorable J. Bennett Johnston, Chairman, Subcommittee on Energy and Water Development, Committee on Appropriations, United States Senate, July 10, 1992.

30 U.S. Department of Defense. "DoD News Briefing — Secretary Rumsfeld and Gen. Myers." May 20, 2003. [http://www.defenselink.mil/transcripts/2003/tr20030520-secdef0207.html]

31 Kathleen Bailey and Robert Barker, "Why the United States Should Unsign the Comprehensive Test Ban Treaty and Resume Nuclear Testing," *Comparative Strategy*, April-June 2003: 134.

32 See Hans Kruger and Edgar Mendelsohn, "Neutralization of Chemical/Biological Ballistic Warheads by Low-Yield Nuclear Interceptors," Lawrence Livermore National Laboratory publication UCRL-ID-110403, August 1992, and Hans Kruger, "Defense Against Biological or Chemical Bomblet Warheads with Nuclear Interceptors," UCRL-ID-123815, March 1996.

science adviser for arms control at the State Department, writes: "Reduced residual radiation weapons would be similar to the 'neutron' bombs, made in the 1970s, with very low fission yield. For classified reasons they cannot be readily made today without compromise to higher priority programs. Suppressing electromagnetic pulse is very difficult because that would require capturing the many high-energy electrons liberated by a nuclear explosion. Ultra-robust warheads could have been developed and tested decades ago, but weren't, perhaps because the nuclear weapons community feared that that could have ended the need to test."[33] The Administration and Congress appear to have little if any interest in nuclear-armed interceptors, as noted earlier.

In a 1989 report, DOE explained its low interest in more robust warheads. It stated that the nuclear weapon laboratories were studying the feasibility of such warheads, which would have reduced sensitivity to aging, to material properties that might change slightly during remanufacture, to tolerances of manufacture, etc. "We cannot predict the success of this research program," DOE said, and noted that "such warheads would be less operationally efficient than the optimized warheads in current use and have some size and weight penalties with attendant cost and performance penalties to weapon delivery systems."[34] Now, though, such penalties are arguably less important than they were during the Cold War — and ultra-robust warheads accordingly more desirable — because the United States plans to reduce the number of warheads carried by operationally deployed ballistic missiles,[35] and bombers can be loaded with fewer bombs and missiles, making extra carrying capacity available on these systems.

Is it Technically Feasible to Use Nuclear Weapons to Destroy Stored Bioweapon Stockpiles? If So, Is Such Use Advisable?

One concept that ACI might explore would be nuclear weapons for destroying chemical or biological warfare agents. ACI supporters advance it as a main reason for studying new or modified weapons, yet there is much less understanding of this topic than of RNEP. This section deals with attacks on biological warfare (BW) agents because some of them, such as anthrax spores,

---

33 Personal communication, September 2, 2003.

34 U.S. Department of Energy. "Annual Report to Congress, Volume I: Program Status of Preparations for Further Limitations on Nuclear Testing, Required by Section 1436 of the National Defense Authorization Act, FY 1989," December 1989, p. 16.

35 U.S. Department of Defense. "Findings of the Nuclear Posture Review," January 9, 2002, slide titled, "NPR Decisions Made."

are more lethal than the most toxic chemical warfare agents on a weight-for-weight basis. It first addresses the technical feasibility of such attacks, drawing heavily on the work of Hans Kruger[36] and Jonathan B. Tucker.[37] Kruger is a physicist and an authority on the effects of nuclear weapons on chemical and biological agents. He led Livermore's weapons effects group from around 1973 to 1999, when he retired, and is currently working at Livermore on these issues. Tucker is a Senior Researcher at the Center for Nonproliferation Studies, Monterey Institute of International Studies. He has written extensively on BW and was a U.N. biological weapons inspector in Iraq.[38] After considering feasibility, this section addresses the advisability of such attacks, taking into account political, intelligence, and collateral-damage issues, and then offers several concluding observations.

Feasibility

In calculating the effects of nuclear weapons on BW agent, Kruger makes several assumptions about the target.[39] The agent would be stacked in steel barrels or larger containers in various configurations, stored in the open or in a typical warehouse. It would be in the form of a water slurry,[40] for which he used water as a surrogate for the purpose of computer modeling.[41] He stated that "[b]iological agent is typically stored in barrels or larger storage containers."[42] He further stated that BW agent in a water slurry is a conservative assumption because less radiation would reach the bottom of a stack of barrels filled with water slurry than with, e.g., dry anthrax powder.[43]

While there may be various ways to neutralize BW agent without the cooperation of the country possessing such agent, Kruger argues that using a nuclear weapon to neutralize BW agents requires applying a lethal dose of radiation to all parts of the storage site. Nuclear weapons produce various forms of radiation. Xrays are stopped near the top of the material because they do not have enough energy to penetrate further. Gamma rays (electromagnetic radiation of higher energy than x-rays) penetrate more deeply. One study finds gamma

---

36 Personal communication, September 16, 19, 23, and 24, and October 7, 2003.

37 Personal communication, September 30, October 1, and October 4, 2003.

38 See also Robert Nelson, "Nuclear 'Bunker Busters' Would More Likely Disperse than Destroy Buried Stockpiles of Biological and Chemical Agents," c. 2003, 20 p., available at [http://www.princeton.edu/~rnelson/papers/agent_defeat.pdf].

39 These assumptions are detailed in H. Kruger, *Radiation-Neutralization of Stored Biological Warfare Agents with Low-Yield Nuclear Warheads,* UCRL-ID-140193, August 21, 2000, p. 1-4. [http://www.llnl.gov/tid/lof/documents/pdf/238391.pdf]

40 Personal communication, October 7, 2003.

41 Kruger, Radiation-Neutralization of Stored Biological Warfare Agents ..., p. 3.

42 Ibid., p. 2.

43 Personal communication, October 7, 2003.

rays of limited value: In two cases studied (attacking BW agent in an underground bunker, one case with a nuclear explosion inside the bunker and the other with the explosion in the ground but outside the bunker), "the amount of bio-agents effectively irradiated by this process [gamma rays] will in all likelihood be a small fraction of the total."[44] Neutrons are the most effective mechanism. They are more penetrating than gammas. In addition, neutrons generate gamma rays when they strike the target material; gammas generated far down in the stack of stored agent can penetrate to the bottom of the stack. The key to calculating the effects of a nuclear weapon on BW agents, then, according to this argument, is the number of neutrons and their energy level; a dose of neutrons and gamma rays sufficient to kill BW agent at the bottom of the stack would expose the rest of the stack to a much greater dose.

The fireball of a nuclear explosion would probably incinerate BW agents, but radiation is a more useful kill mechanism from a military point of view because it can be calculated much more precisely if one knows the conditions under which the agent is stored. Currently-deployed nuclear weapons, which use a combination of fission and fusion to produce an explosion, must be detonated close to the ground — creating much fallout — in order to kill BW agents; for a greater height of burst, the number and energy level of neutrons produced would be insufficient for this task.[45] Nuclear weapons can be made, however, that generate more neutrons. Such a weapon is a reduced blast/enhanced radiation weapon, such as the "neutron bomb" of the 1970s. (Neutron bombs were controversial and are discussed further below.) For a given yield, these weapons produce an order of magnitude more neutrons than do standard low-yield nuclear weapons, and the neutrons have a higher energy level. Kruger calculates that a 10-kt weapon of this type burst at a height of 10 meters would neutralize BW agents to a radius of about 50 meters, vs. 10 meters for a 10-kt fission

[44] Michael May and Zachary Haldeman, "Effectiveness of Nuclear Weapons against Buried Biological Agents," May 2003, p.14. Available at [http://www.ciaonet.org/wps/mam09/mam09.pdf].

[45] Kruger provided the following details: "most of the fallout is from the radioactive fission products (the two pieces into which a uranium or plutonium atom splits when it fissions) that is initially vaporized and then solidifies on the solid matter 'scooped up' from the surface. ... think of it as a vapor plating type of process. This results in relatively large particulates that drop sufficiently fast to reach the ground within a few to a few tens of kilometers from ground zero. The same amount of radioactive fission products is present when the detonation occurs at such a height that no ground material is scooped up. However, now the fission products form much smaller particles that have a much lower settling velocity. Now the fission debris is spread over a very much larger distance and as a result the radioactive dose on the ground is very much lower." Personal correspondence, September 24, 2003.

weapon at the same height of burst.[46] (For comparison, the Hiroshima bomb had a yield of 15 kt.) He notes several consequences of the increased neutron flux:

> a fusion warhead [e.g., a neutron bomb] has a larger sterilization area than a fission warhead of equal yield. Another (to me: significant) consequence is that I can detonate a fusion warhead at an altitude above the ground (higher than the "optimum" altitude) sufficiently high that the fireball does not touch the ground, and still produce a sterilizing dose at the very bottom of the barrel stack over a militarily significant area. Avoiding fireball contact with the ground eliminates most of the local fallout. This is not feasible for a fission warhead.[47]

If an adversary stored BW agent in an underground bunker, the surrounding earth, steel, and reinforced concrete would significantly attenuate the neutrons from a weapon that burst far enough from the bunker (vertically or horizontally) that the fireball did not enter it. To destroy the agent, an earth penetrator weapon (EPW) would have to place the fireball inside the bunker. The chief kill mechanism would be the heat of the fireball, the effectiveness of which would be increased because the fireball would (at first) be concentrated within the structure. The delayed radiation resulting from the fission debris would expose any remaining agent to a further radiation dose.[48]

Critics question Kruger's postulated target as optimized for destruction by a nuclear weapon. The next few paragraphs focus on anthrax because it is one of the tougher BW agents to kill and is also one of the most militarily useful agents.

Would a nation make anthrax in aqueous form? It is easier to make anthrax as a water-based slurry than a fine powder. According to Tucker, "It appears that at the time of the 1991 Gulf War, Iraq only had the capability to produce wet anthrax. Iraqi BW scientists appear to have worked during the 1990's to acquire the know-how to produce dry, powdered anthrax, using a closely related bacterium (the biopesticide Bt) as a model system. If a country had the capability to produce dried, powdered anthrax, it would presumably store it in that form because of its much longer shelf-life and the fact that dry, powdered anthrax is much easier to disseminate as an aerosol from an aircraft sprayer system or a missile warhead." An Office of Technology Assessment (OTA) report of 1993 provides further information: "when anthrax bacteria are incubated under particular conditions, they transform themselves into the rugged

---

[46] Kruger, Radiation-Neutralization of Stored Biological Warfare Agents..., p. 7.

[47] Personal communication, September 16, 2003.

[48] For a detailed technical analysis of the mechanisms by which a nuclear explosion would destroy BW agent stored underground, see May and Haldeman, *Effectiveness of Nuclear Weapons against Buried Biological Agents.*

spore form, which has a long shelf-life ... This spore-forming ability makes anthrax particularly well suited for delivery by missiles or bombs. The spores are stable when suspended in air, can survive explosive dissemination from a bomb or shell, and ... will live for several days if direct sunlight is avoided."[49]

Would a nation need huge quantities of anthrax? The OTA report states that "a gram of anthrax [in spore form] theoretically contains some 10 million lethal doses."[50] The ability of a single envelope with a small quantity of powdered anthrax to shut down the Capitol complex in October 2001 is a case in point. Tucker states, "A bioterrorist attack in a confined space, such as a subway station, would require a few ounces of dry anthrax. Military applications or strategic attacks against cities would require much more material, perhaps hundreds of pounds to attack a large city. As a rough rule of thumb, about a kilogram of dry, concentrated anthrax spores would be required to infect 50 percent of the people in a square kilometer, assuming efficient aerosol dissemination under optimal conditions. A kilo of dry anthrax is equivalent to a much larger volume of wet slurry."

Would a nation make and store anthrax long before it intended to use it? According to Tucker, "A nation would probably not make anthrax far in advance of use unless it had mastered the technology to produce dry powdered anthrax, which has a shelf life of several years. A wet slurry of anthrax will tend to coagulate after several months, rendering it unusable as a weapon. Thus, a country that possessed only wet-anthrax technology would probably produce the agent shortly before use."

Would a nation store large quantities of anthrax aboveground? Attacking anthrax stored in shallow underground bunkers would require the use of EPWs to explode in the bunker; otherwise, the ground would shield the bunker from neutrons.[51] An EPW would create much fallout and could cause other problems, as detailed in the section on RNEP, below. Storing anthrax in a bunker at a depth considerably greater than that which EPWs could reach would prevent neutrons, heat, and delayed radiation (from radioactive material left by the nuclear explosion) from reaching the bunker. Tucker states that "[s]ince anthrax spores are quite rugged and are in a state of suspended animation, it's reasonable

---

49 U.S. Congress. Office of Technology Assessment. *Technologies Underlying Weapons of Mass Destruction.* OTA-BP-ISC-115. USGPO, December 1993, p. 79.

50 Ibid., p. 78.

51 For a detailed technical analysis of the mechanisms by which a nuclear explosion would destroy BW agent stored underground, see Michael May and Zachary Haldeman, "Effectiveness of Nuclear Weapons against Buried Biological Agents," Center for International Security and Cooperation, Stanford University, May 2003, 27 p.: [http://iis-db.stanford.edu/pubs/20216/May_Haldeman_final.pdf].

to assume that [ground] shock alone would be unlikely to kill them reliably." The statement from the OTA report that anthrax spores can survive explosive dissemination, cited above, supports this point. For further protection from neutrons, an adversary could shield anthrax with material that reflects neutrons (e.g., beryllium) or that absorbs them (e.g., polyethylene mixed with boron). Alternatively, dry powdered anthrax could be hidden anywhere in small lots, in which case its destruction would depend mainly on locating the agent; once that is done, various types of weapons and tactics might be used to neutralize it.

Advisability

Critics point to a number of intelligence hurdles. A successful attack would require detailed knowledge of the target, such as storage configuration. Knowing the target's precise location would be crucial because of the very short range at which even an enhanced radiation weapon would sterilize BW agent. Yet as Tucker notes, "During the 1991 Gulf War, many suspect BW production and storage facilities were targeted in error, and several real facilities such as the al-Hakam factory — which UNSCOM[52] later determined was Iraq's largest anthrax production facility — were totally unknown to U.S. intelligence." It would be important to know the type of agent because, as Kruger states, "the neutralizing dose [of radiation] for different bio agents varies by a couple of factors of two." Compensating for uncertainty as to which agent was being attacked by using a weapon able to neutralize the most radiation-resistant agent could result in using a weapon with a yield several times larger than needed.

Critics also note political problems with nuclear weapon development and use. An attack that sought to minimize fallout would have to use reduced blast/enhanced radiation weapons. Neutron bombs, which the United States developed in the 1970s, received intense domestic and foreign criticism as the "capitalist bomb" that spared property while killing people. Similar criticism would doubtless reemerge if the United States began a program to develop such weapons, especially if the intended targets were perceived to be rogue states in the Third World. The development program could require nuclear testing, fueling further criticism.

Critics assert that there are nonnuclear approaches to destroying BW agents. According to one report, the Defense Department is working on nonnuclear weapons for destroying chemical or biological agents.

---

[52] The U.N. established UNSCOM, the United Nations Special Commission, in 1991 to supervise the elimination of chemical and biological weapons and certain missiles in Iraq following the Gulf War. See [http://www.un.org/Depts/unscom/] for details.

> Destroying the actual chemical or biological material [in Iraq] is a task that may fall to agent defeat weapons being developed by the U.S. Navy and Lockheed Martin under a program originally dubbed Vulcan Fire and now spearheaded by the secretive U.S. Defense Threat Reduction Agency. The HTI-J-1000, as it is called, would be the fill inside the penetrating warhead used on the massive 2,000-pound GBU-24 laser-guided bomb and BLU-109 Joint Direct Attack Munition (JDAM) used to attack underground bunkers.[53]

That report states that the chemical reactions would generate intense heat that would destroy the biological and chemical agents, and would generate other chemicals that would neutralize these agents.[54]

Michael Levi, director of the Strategic Studies Project at the Federation of American Scientists, suggested another approach to defeating a hypothetical underground anthrax storage bunker in Iraq:

> U.S. planners may not want to directly attack the bunker. Instead, a watch could be placed on the facility using satellite imagery coupled with armed unmanned aerial vehicles. Anyone or anything attempting to enter or leave the bunker would be destroyed, making the anthrax inside unusable.[55]

Kruger rejects the idea of using conventional high explosives to neutralize BW agents for two reasons. First, he calculates that it would take a huge quantity of these explosives, perhaps hundreds of 500-pound bombs, to raise the temperature of the quantity of agent in a typical storage site sufficiently, assuming the agent was in a water slurry. Similarly, if acid were used instead of explosives to kill BW agent, it would take a very large quantity of acid. Second, he argues that attacking a BW storage site with conventional weapons would cause much more collateral damage than would an attack with nuclear weapons, even standard low-yield weapons:

> Conventional weapons are likely to only kill a small fraction of the agent while dispersing a significant fraction of the remaining live agent. My past dispersal calculations for a generic target in a country of current interest have shown collateral damage areas from dispersed bio agent that were very many orders-ofmagnitude larger than the fallout area for the ground burst of a low-yield fission warhead.[56]

---

[53] Frank Vizard, "Extinguishing the Threat," *Scientific American.com,* February 18, 2003. http://www.scientificamerican.com/article.cfm?articleID=00000CB6-18E9-1E4D-967D8 09EC588EEDF&catID=4

[54] Ibid.

[55] Michael Levi, "The Case against New Nuclear Weapons," *Issues in Science and Technology,* Spring 2003: 65-66.

[56] Personal communication, September 16, 2003.

It might be possible to deploy reduced blast/enhanced radiation weapons without testing them, such as by adapting designs made and tested several decades ago. Optimizing these weapons for defeating specific BW agents, however, could involve changes that would eventually lead to testing.[57]

An attempt to seal off a facility encounters at least two problems. First, a U.S. attack on a rogue state with no warning would be unlikely. With warning, that state could easily move materials and equipment to other sites before the attack. Indeed, leaders who believed that their nation was at risk from a U.S. attack would probably make plans well in advance to do just that. Second, some nations, such as Afghanistan and North Korea, have extensive tunnel networks that might provide undetected access, at a distance, to military facilities connected to such networks.

Conclusion

It appears that, under certain specific circumstances, a nuclear blast is technically capable of sterilizing BW agents. One example would be a large quantity of agent in aqueous solution stored at or near the Earth's surface, with the location pinpointed within meters and with accurate intelligence as to the agent present and how it is stored. (The agent would have to be stored long enough that intelligence assets could locate it, but not so long that it would deteriorate significantly.) Another example would be a depot with a ton of powdered anthrax, in which case a relatively small amount (as compared with hundreds of barrels) of agent could do immense harm and it would be imperative to minimize the risk of anthrax escaping. Few would suggest, though, that nuclear weapons would automatically be the means of first choice for attacking BW agent. Nuclear-weapon use entails clear political disadvantages, the effectiveness of nuclear weapons in particular situations may be uncertain, and nonnuclear means may be available. Whatever the case, it appears likely that Congress would examine closely any request to proceed to development engineering or production of nuclear weapons intended to destroy BW agents. Such examination might include the following questions:

How probable is the military scenario put forth to justify development and acquisition of these weapons?

How plausible is the scenario? What are its flaws? How broad or narrow are the circumstances under which the scenario exists?

Would nuclear use be the only option in that scenario, or could nonnuclear means suffice? What is the state of progress on nonnuclear alternatives?

---

[57] Information provided by Livermore staff, September 25, 2003.

What are the military and political pros and cons of developing and acquiring these weapons?

Can ACI Help in the Fight Against Nuclear Proliferation?

According to the FY2003 report of the Foster Panel,

> A second requirement defined in the NPR is for advanced concept exploration. Work must proceed on future concepts and technologies to avoid technical surprises, to attract and train future stockpile stewards, and to assess intelligence information on the continuing development and proliferation of WMD and their delivery methods. The Panel emphasizes that it is essential for Congress to be kept apprised of developments in foreign weapons programs and their potential implications for U.S. security.[58]

ACI could improve U.S. understanding of potential terrorist nuclear weapons, which could improve U.S. ability to interpret fragments of intelligence, to detect evidence of nations or groups working to develop nuclear weapons, to detect the transport of such weapons and attempts to bring them into the United States, and to disable such weapons. It could also help avoid technical surprise by improving U.S. ability to understand foreign weapon developments. For example, creating nuclear weapon designs consistent with the Chinese or North Korean weapons program may help the United States interpret intelligence clues to gain insight into the pace of the program, its technological successes and failures, and warhead characteristics.

ACI's supporters reject the claim that the program will encourage nuclear proliferation. They argue that nations will decide to go nuclear for reasons far larger than ACI. Even if ACI encouraged proliferation, that would not necessarily mean that the United States should halt the program. It is likely, supporters claim, that North Korea and Iran are pursuing nuclear weapons out of fear of U.S. conventional forces. Does this mean, supporters ask, that the United States should halt military modernization and shrink its conventional forces to keep those nations from going nuclear? At the same time, it is argued, India and Pakistan are pursuing nuclear weapons not because of U.S. conventional or nuclear forces but because of threats that each sees from the other, and that India sees from China; these dynamics have nothing to do with ACI.

ACI's critics maintain that designing new nuclear weapons tailored for use against targets in nonnuclear weapon states would undercut U.S. nuclear

---

[58] FY 2003 Report to Congress of the Panel to Assess the Reliability, Safety, and Security of the United States Nuclear Stockpile, p. 3.

nonproliferation efforts by in effect telling the world to do what we say, not what we do.

These critics assert that the program is scarcely needed to improve U.S. ability to understand nuclear weapons developments made by rogue states or terrorists. U.S. nuclear weapons experience accumulated over the past six decades provides ample information for interpreting intelligence clues connected with the design and fabrication of such weapons. Nor is ACI needed to improve U.S. ability to understand Russian or Chinese nuclear weapons programs. The Stockpile Stewardship Program has greatly increased U.S. weapons knowledge, arguably placing this nation in a better position now to interpret intelligence data on Russian and Chinese nuclear programs than it was a decade ago.

Another argument is that rogue states want nuclear weapons to deter U.S. nuclear forces. A North Korean statement of August 2003 makes this point:

> The Bush administration openly disclosed its attempt to use nuclear weapons after listing the DPRK as part of 'an axis of evil' and a target of a 'preemptive nuclear attack.'
> This prompted us to judge that the Bush administration is going to stifle our system by force and decide to build a strong deterrent force to cope with it. ... Our deterrent force ... is a means for self-defence to protect our sovereignty.[59]

Thus, it is argued, nuclear weapons designed to threaten key targets in rogue states will only accelerate development of nuclear weapons by those states.

Does ACI Offer Unique Value for the Nuclear Weapons Enterprise?

ACI's proponents believe that the United States must retain weapons design capability because new weapons might be needed in the future, such as to attack specific targets or to replace current weapons that deteriorate beyond the point of repair. To that end, it is further argued, the United States should avoid a situation in which it needs a new weapon yet no one at the labs has had the experience of designing one.

Supporters see ACI as complementing the Stockpile Stewardship Program. Maintaining existing weapons involves predicting, detecting, and fixing

---

[59] Kim Yong Il, Vice Minister of Foreign Affairs for Democratic People's Republic of Korea, text of speech delivered at six-country nuclear talks in Beijing, as published by KCNA, the DPRK's official news agency; posted by Reuters, August 29, 2003. [http://www.cnn.com/2003/WORLD/asiapcf/east/08/29/nkorea.text.reut/index.html]

problems with existing weapons. Designing new weapons, however, arguably involves different skills and thought processes, just as editing differs from writing. A designer might be given a target and required warhead characteristics (yield, weight, compatibility with a specific delivery vehicle, etc.) and be tasked to design a warhead that produces the required military effectiveness while minimizing collateral damage. Experienced weapon designers from all three labs have stated to NNSA headquarters that the current Stockpile Stewardship Program (maintenance of the enduring stockpile) is not sufficient to provide the range of opportunities necessary to train and mentor the next generation of designers.[60]

ACI, it is argued, will provide such training. The purportedly leaked NPR stated that ACI "will provide unique opportunities to train our next generation of weapon designers and engineers."[61] Los Alamos Director John Browne stated that ACI

> articulates a strategy for ... transferring nuclear warhead design knowledge, and exercising design skills. This initiative provides an outstanding opportunity for the nuclear weapons complex to ensure that existing expertise is transferred to a future generation of stockpile stewards, and to extend the front-line weapons lifetimes beyond that of the designers who designed and tested them.[62]

ACI will also develop new links between the weapons community and DOD agencies. It is helpful, proponents believe, for some members of the weapons community to have personal contacts with members of the U.S. Strategic Command and the Defense Threat Reduction Agency, among others, to gain an understanding of what weapons might be of value to DOD, and to give DOD a sense of what weapons technologies may be available.

The harshest critics of nuclear weapons believe that there is no need to keep alive the skills to design new weapons. Current weapons, they argue, are more than sufficient, and their designs have been proven. Until nuclear weapons are abolished, these critics would maintain the existing stockpile; in their view, the multibillion-dollar stockpile stewardship program is excessive for this purpose. This school of thought would reject ACI completely.

Some critics of ACI see a value in maintaining design skills so as to improve U.S. ability to understand warhead problems that may emerge, to

---

60 Personal correspondence, NNSA official, September 3, 2003.

61 Nuclear Posture Review [Excerpts], p. 35.

62 "Statement of John C. Browne, Director, Los Alamos National Laboratory, Submitted to United States House of Representatives, Committee on Armed Services, Subcommittee on Military Procurement, June 12, 2002." [http://www.lanl.gov/orgs/pa/News/testimony 061202.shtml]

understand how changes made during refurbishment may affect weapon performance, and to design replacement weapons should that be necessary. Such skills can be maintained, they would argue, by designing weapons similar to current ones rather than designing new types of weapons. That approach would not require ACI.

Critics would question training as a justification for ACI. The number of people trained by ACI is very small, and there are already training programs for junior staff. In October 1998, Sandia started an Intern Program, a two-year course for new employees on the nuclear weapons program and Sandia's role in it. In October 1996, Los Alamos started the Theoretical Institute for Thermonuclear and Nuclear Studies on nuclear weapons design for new personnel in that field.[63]

Some question whether ACI, as proposed for FY2004, offers a useful plan for maintaining weapon design capabilities. For example, Raymond Jeanloz, Professor of Earth and Planetary Science at University of California, Berkeley, and a member of the University of California President's Council,[64] argues that repackaging an existing bomb with no work on the nuclear explosive component (RNEP), working on low-yield designs (an area extensively studied in the past, including with underground tests), or other ideas proposed for ACI amount more to reworking old ground than to significantly exercising design activity.[65]

There is no need for new links between DOD and the nuclear weapons establishment, in the critics' view, because several DOD-DOE groups have been in existence for many years. These include the Nuclear Weapons Council, program officer groups, and the Stockpile Assessment Team of Strategic Command's Strategic Advisory Group.

---

[63] U.S. Commission on Maintaining United States Nuclear Weapons Expertise. *Report to Congress and the Secretary of Energy*, p. 20.

[64] The University of California operates three DOE laboratories: Los Alamos, Lawrence Livermore, and Lawrence Berkeley National Laboratories. This council provides oversight of these labs for the university.

[65] Personal correspondence, September 2, 2003.

Chapter 5

ROBUST NUCLEAR EARTH PENETRATOR

DESCRIPTION

For FY2004, the Administration requested $15.0 million to continue a study, which began in May 2003, of the Robust Nuclear Earth Penetrator, or RNEP. The NDAA provided the amount requested, while the EWDAA provided $7.5 million. The FY2005 request is $27.6 million.

The RNEP study, part of the Advanced Concepts Initiative, is examining the desirability and feasibility of modifying an existing B61 or B83 bomb to enable it to penetrate into the ground before detonating. By so doing, a weapon of specified yield would have a much greater effect against hardened and deeply buried targets — such as are used in some rogue states to protect key assets — than would a nuclear weapon of the same yield burst on the Earth's surface. Modification would mainly involve strengthening the weapon's case, rather than changing the nuclear explosive package. In the 1990s, DOE conducted a similar modification, converting some B61 bombs into B61-11 earth penetrators, a modification that did not require nuclear testing. If NNSA converts B61s or B83s into RNEPs, current plans envision that the conversion would not require nuclear testing.

While the histories of earth penetrator weapons and the Spratt-Furse ban on lowyield R&D are intertwined, repeal of that ban does not necessarily open the door to RNEP. The ban applied only to weapons under 5 kt; RNEP, if it proceeds, would be a modification of the B61 bomb or the B83 bomb, both of which are available in a number of yields, with the maximum said to be far above 5 kt. Further, both have been in the stockpile for many years. Even if they had yields only below 5 kt, the ban applied only to weapons that had not entered production by November 30, 1993.[1]

Technical Background

Structures can be hardened and deeply buried in an attempt to withstand an attack by conventional munitions. Such structures may be used to protect leadership, biological agent production facilities, nuclear weapon storage sites, and other strategic targets. Reports reference such facilities in Iraq, North Korea, and Iran.[2] A DOD-DOE report indicates that thousands of such structures exist worldwide, many in nations that do not pose a threat to the United States.[3]

While special operations forces or precision-guided conventional bombs might defeat deeply buried structures by attacking power supplies, ventilation systems, and exits, these structures would probably have backup power supplies, escape tunnels, etc. Accordingly, destruction would be a surer means of completely eliminating the threat they pose. Facilities at shallow depths, perhaps 10 meters, are vulnerable to destruction by conventional munitions that penetrate the ground. Destruction of facilities buried at greater depths may require a nuclear weapon. During the Cold War, the United States deployed the B53 nuclear bomb, a weapon said to have a yield of some megatons (thousands of kilotons), for such missions. A weapon of such high yield would create blast and fallout that would destroy surface structures and kill inhabitants out to many miles; in the current environment, a President would be unlikely to order its use against a target in a rogue state so as to avoid killing very large numbers of civilians. As C. Paul Robinson, President of Sandia National Laboratories, said, we would be self-deterred from using high-yield weapons.[4]

---

1 The definition of a new weapon in P.L. 107-314, Section 3143, cited above, appears to preclude the argument that modifications of the B61 or B83 would constitute a "new" weapon.

2 Glenn Kessler, "Nuclear Sites in Iran Worry U.S. Officials," *Washington Post,* December 14, 2002: 18, 21; David Fulghum, "Iraq's Hidden Weapons 'Are Likely Underground,'" *Aviation Week & Space Technology,* December 16, 2002: 44; and Joby Warrick, "Iran, N. Korea Nuclear Plans Pose New Risks," *Washington Post,* December 24, 2002: 1, 11.

3 Departments of Defense and Energy. Report to Congress on the Defeat of Hard and Deeply Buried Targets ..., p. 8.

4 C. Paul Robinson, "Maintaining a Viable Nuclear Weapons Program in a Test Ban Environment: A Strong Technical Foundation in the Laboratories," address presented at the Nuclear Security Decisionmakers Forum, March 28, 2000, Albuquerque, NM: [http://www.sandia.gov/media/speeches/NSDMF2000.doc]

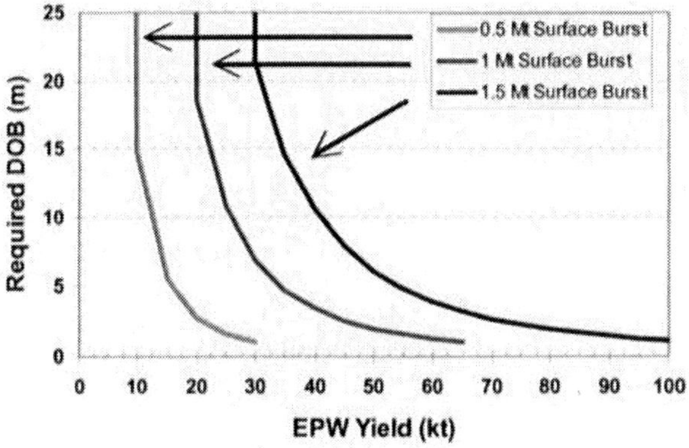

Figure 1. Earth penetration reduces yield needed to destroy buried targets. Source: Steven Hatch, Dandia National Laboratories, 2000. Note: Mt, Megaton; EPW, earth penetrator weapon; DOB, depth of burst .

At issue, insofar as the nuclear option is concerned, is how to destroy hardened and deeply buried facilities with lower-yield weapons. Much of the energy of a ground-burst weapon is reflected back into the air. Scientists on both sides of the larger debate agree that a nuclear weapon that penetrates even a short distance into the ground will transfer much more of its energy into ground shock, which can destroy buried targets. For example, Sandia reports that a 1-megaton (1,000-kiloton) weapon burst on the Earth's surface has about the same effect on buried targets as a 33-kiloton weapon detonated at 5 meters depth. Most of the payoff comes from burying the weapon several meters in the ground, especially for lower-yield weapons; burial beyond several meters produces diminishing returns.[5] Figure 1 illustrates these points. Robert Nelson reports similar data.[6] Thus a lower-yield earth penetrator weapon (EPW) can have the same effect on buried targets as a much higher-yield surface-burst weapon. This large reduction in yield, in turn, reduces fallout and blast damage. A common misconception is that detonating a weapon a few meters underground reduces fallout; in fact, a weapon of specified yield detonated at shallow depth in wet soil will create more fallout than if it were detonated on the surface. The reduction in fallout from lower yield, though, can offset the increase in fallout from a shallow subsurface burst. Continuing the example just noted, fallout with a specific intensity (150 rads in 24 hours for an unsheltered

[5] Memorandum, "Comparison of Surface and Sub-Surface Nuclear Bursts," from Steven Hatch, Sandia National Laboratories, to Jonathan Medalia, October 30, 2000.

[6] Robert Nelson, "Low-Yield Earth-Penetrating Nuclear Weapons," *Science and Global Security,* 10: 3-4, 2002.

person) would, by one calculation, be generated over 2,700 square kilometers by a 1-megaton surface-burst weapon, or over 150 square kilometers by a 33-kiloton weapon detonated at a depth of 5 meters.[7] Figure 2 illustrates these points (though it assumes individuals would be sheltered).

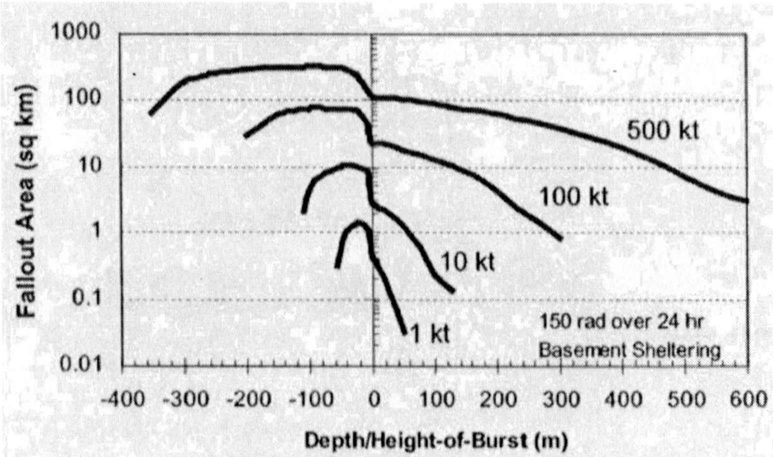

Figure 2. Lower yield reduces fallout; shallow depth of burst increases it. Source: Steven Hatch, Dandia National Laboratories, 2000.

Lower-yield weapons, even EPWs, may not suffice to destroy all buried targets. Stephen Younger, then Associate Laboratory Director for Nuclear Weapons, Los Alamos National Laboratory, wrote, "Some very hard targets require high yield to destroy them. No application of conventional explosives or even lower-yield nuclear explosives will destroy such targets, which might include hardened structures buried beneath hundreds of feet of earth or rock." Indeed, "Superhard targets, such as those found under certain Russian mountains, may not be able to be defeated reliably by even high-yield nuclear weapons."[8] Similarly, Robert Nelson wrote, "A low-yield nuclear EPW would still only be able to destroy facilities relatively close to the surface ... Very large

[7] Personal communication from Steven Hatch, Sandia National Laboratories, September 4, 2003. The level of fallout intensity used is 150 rad over 24 hours for a person in a basement shelter. (A dose of 100-200 rem produces minimal symptoms in humans, while a dose of 300-600 rem may kill half of those exposed. Rads and rems are measures of energy absorbed by the body; for most (but not all) forms of radiation, one rad equals one rem. Dale Moeller, *Environmental Health,* Cambridge, Harvard University Press, 1997, p. 249-250.)

[8] Stephen Younger, "Nuclear Weapons in the Twenty-First Century," Los Alamos National Laboratory, LAUR-00-2850, June 27, 2000, p. 10.

yield (>~100 kt) weapons are still required to destroy facilities buried under the equivalent of 100 m of concrete."[9]

Even low-yield EPWs would generate much fallout. For example, photographs of two nuclear tests show the fallout cloud generated by low-yield weapons buried hundreds of feet with the emplacement hole sealed. A test named "Des Moines,"a 2.9-kiloton test conducted June 13, 1962, at the Nevada Test Site, was buried in a mountain. Pictures show a huge cloud of radioactive dust, with cars and pickups of spectators rushing away from it. Another test, "Baneberry," with a yield of 10 kt, was conducted December 18, 1970, in a vertical shaft some 900 feet deep at the Nevada Test Site.[10] The fallout cloud rose about 10,000 feet into the air,[11] and was tracked to the Canadian border.[12] According to DOE, "Following Baneberry, new containment procedures were adopted to prevent a similar occurrence."[13] The minimum depth of burial for even the smallest test is 600 feet.[14] An EPW, in contrast, would be buried perhaps a few tens of feet and the path by which it penetrated the earth would not be sealed. Fallout from such weapons would not be contained at all, and standard-yield EPWs under consideration would produce more fallout than low-yield EPWs.

Improvements in accuracy also permit a reduction in warhead yield for attacking buried targets. Four Los Alamos scientists calculate that a moderately hard structure buried at shallow depth could be destroyed by a 15-kiloton weapon with an accuracy of 100 meters, or by a 0.5-kiloton weapon with an accuracy of 5 meters, with both weapons burst on the surface. Earth penetration further reduces the yield needed. The combination of improved accuracy and earth penetration, they argue, sharply reduces the yield required and the collateral damage.[15]

---

[9] Nelson, "Low-Yield Earth-Penetrating Nuclear Weapons," p. 18.

[10] The yield of these tests is from U.S. Department of Energy. Nevada Operations Office. Office of External Affairs. *United States Nuclear Tests: July 1945 Through September 1992.* DOE/NV-209 (Rev. 14), December 1994, p. 14, 41.

[11] U.S. Department of Energy. Nevada Operations Office website: [http://www.nv.doe.gov/news&pubs/photos&films/general.htm]. This site also provides depth of burial and a photograph of the test.

[12] U.S. Congress. Office of Technology Assessment. *The Containment of Underground Nuclear Explosions.* OTA-ISC-414, October 1989, p. 31.

[13] U.S. Department of Energy. Nevada Operations Office website: [http://www.nv.doe.gov/news&pubs/photos&films/general.htm]

[14] Office of Technology Assessment, The Containment of Underground Nuclear Explosions, p. 36.

[15] Bryan Fearey, Paul White, John St. Ledger, and John Immele, "An Analysis of Reduced Collateral Damage Nuclear Weapons," *Comparative Strategy,* October/November 2003, p. 314-315.

While the foregoing discussion has focused on technical aspects of nuclear weapon effects, it must be kept in mind that even low-yield earth penetrator weapons could, depending on the circumstances, kill large numbers of people. The calculation for such a weapon burst near Damascus, cited above, illustrates this point.

HISTORY THROUGH THE FY2003 BUDGET CYCLE

Interest in destroying buried targets traces back to the early days of the Cold War when the Soviet Union first built underground command posts and missile silos. Current interest in EPWs for use against HDBTs traces back to the Persian Gulf War, as is detailed in the earlier section on low-yield R&D.

This latter experience led to the development of the current EPW, the B61 mod 11, or B61-11. (B61 is the basic weapon, while the mod number indicates a modification, converting the bomb to a specific purpose, delivery by a particular aircraft, etc.) The modification involved a strong case and some internal strengthening. Because it entailed only minimal changes to the nuclear explosive package, DOE did not conduct nuclear tests of the weapon.[16] DOE converted a small number of older B61-7 bombs to B61-11s in the mid- to late 1990s. The B61-11, however, has its limitations. The Nuclear Posture Review stated, according to purportedly leaked excerpts, that the weapon "has a very limited ground penetration capability" and "cannot survive penetration into many types of terrain in which hardened underground facilities are located."[17]

The growth in number of buried targets and the weaknesses of the B61-11 led DOD and DOE to study RNEP. According to a U.S. Strategic Command officer,

> Our current arsenal, developed in the Cold War, was not designed to address this growing worldwide threat [of buried targets]. There are facilities today which we either cannot defeat, even with existing nuclear weapons, or must hold at risk using a large number of weapons. As a result, both the Department of Defense and the Department of Energy, through the Nuclear Weapons Council, have approved a study of how to effectively counter this threat. This

---

[16] According to a Los Alamos publication, "The essence of the modification is a field changeout of the weapon's case to provide an earth-penetration capability." "B61-11 Update," *Weapons Insider,* August 1996: 2. Further, "modifications to the warhead's electrical system and physics package were minimal," "B61-11 NESE Completed," *Weapons Insider,* June/July/August 1998: 5.

[17] Nuclear Posture Review [Excerpts], p. 47.

> study of a Robust Nuclear Earth Penetrator (RNEP) will evaluate modifications to existing nuclear weapons that do not require nuclear testing.[18]

Linton Brooks, then Acting Administrator of NNSA, provided further details on the study:

> this study will examine whether or not two existing warheads in the stockpile —the B61 and the B83 — can be sufficiently hardened through case modifications and other work to allow the weapons to survive penetration into various geologies before detonating. This would enhance the Nation's ability to hold hard and deeply buried targets at risk.[19]

Thus RNEP would not necessarily be a low-yield weapon. John Gordon, then Director of the National Nuclear Security Administration, stated in 2002 that the emphasis is on "a more standard yield system called an enhanced penetrator ...There's no design work going on low-yield nuclear weapons."[20] The cited advantage of the higher yield is that while EPWs can destroy a number of buried targets with less yield than a surface-burst weapon, increasing yield increases the radius of damage to buried targets, some of which may be too hard, or too deep, to destroy with a low-yield weapon. In July 2003, NNSA provided further detail on this topic. The ongoing RNEP study will select one of the two candidate bombs. The study aims to retain all existing capabilities. It is not part of the study to create new, lower yields.[21]

NNSA had planned to begin the RNEP study in FY2002 by redirecting $10 million from Directed Stockpile Work Research and Development, a program within stockpile stewardship.[22] As discussed below, however, FY2003 legislation had the effect of delaying this study into FY2003. For FY2003, the Administration requested funds to begin this RNEP study. The planned budget

---

[18] "Statement of RADM John T. Byrd, USN, Director of Plans and Policy, United States Strategic Command, before the House Armed Services Committee, Procurement Subcommittee, June 12, 2002." The Joint Nuclear Weapons Council is a small DOD-DOE agency established by 10 USC 179 that coordinates nuclear weapons activities between the two departments. Statement is available at: http://armedservices.house.gov/openingstatementsandpressreleases/107thcongress/02-06-12byrd.html

[19] "Statement of Linton F. Brooks, Acting Under Secretary of Energy and Administrator for National Security, National Nuclear Security Administration, U. S. Department of Energy Before the Subcommittee on Strategic Forces, Committee on Armed Services, U. S. Senate, April 8, 2003": [http://www.senate.gov/~armed_services/statemnt/2003/April/Brooks.pdf]

[20] U.S. Congress. Senate. Committee on Armed Services. Hearing: Nuclear Posture Review. February 14, 2002, n.p. Transcript prepared by eMediaMillWorks, Inc.

[21] Information provided by NNSA staff, July 25, 2003.

[22] Information provided by NNSA staff, August 11, 2003. See also Walter Pincus, "Nuclear Warhead Study Aims at Buried Targets," *Washington Post,* March 15, 2002: 16.

for the study was $15.0 million a year for FY2003-FY2005, inclusive. Brooks said, "The RNEP feasibility and cost study is currently scheduled for completion in 2006; however, we are looking at opportunities to reduce study time."[23]

The House Armed Services Committee supported the FY2003 request. It included a provision calling for a National Academy of Sciences study on the shortand long-term effects on civilians and military forces of an EPW, an above-ground nuclear explosion to destroy buried targets, and use of conventional weapons to destroy WMD storage or production facilities.[24] Representative Markey offered an amendment on the House floor to bar permanently the use of funds to develop or test a nuclear EPW, and to bar FY2003 funds for conducting a feasibility study of that weapon. The amendment was defeated, 172-243, on May 9, 2002.[25]

The Senate Armed Services Committee recommended eliminating funds for RNEP and requiring the Secretaries of Defense and Energy to report on RNEP, including military requirements, employment policy, targets, and ability of conventional weapons to "address" types of targets that RNEP would hold at risk.[26] The conference bill included the House provision for a National Academy of Sciences study; as of August 2003, it appears that the study will be completed around April 2004.[27] The conference bill fully funded RNEP but barred obligation of FY2003 funds for it until 30 days after the report called for by the Senate Armed Services Committee had been submitted to the Armed Services Committees. DOD submitted that report in classified form on March 19, 2003; the study began on May 1, 2003, with a meeting of NNSA, DOD, and nuclear weapons laboratory personnel in Washington. The FY2003 Energy and Water Development Appropriations Act included the amount requested for Stockpile R&D, the category within Directed Stockpile Work that includes RNEP.[28] NNSA indicated that this measure provided the full $15.0 million requested for RNEP.[29]

---

[23] Statement of Linton F. Brooks, April 8, 2003.

[24] U.S. Congress. House. Committee on Armed Services. *Bob Stump National Defense Authorization Act for Fiscal Year 2003.* H.Rept. 107-436, 107th Congress, 2nd Session, USGPO, 2002, p. 343-344.

[25] U.S. Congress. *Congressional Record,* May 9, 2002: H2339-H2343.

[26] U.S. Congress. Senate. Committee on Armed Services. *National Defense Authorization Act for Fiscal Year 2003.* S.Rept. 107-151, 107th Congress, 2nd Session, USGPO, 2002, p. 473.

[27] Information provided by NNSA staff, August 11, 2003.

[28] Conference Report on H.J.Res. 2, Consolidated Appropriations Resolution, 2003. In U.S. Congress. *Congressional Record,* February 12, 2003: H1035.

[29] Information provided by NNSA staff, August 1, 2003.

LEGISLATIVE ACTIONS IN THE FY2004 BUDGET CYCLE

For FY2004, the Armed Services Committees and their respective Houses included the full $15.0 million request in the defense authorization bills. The House rejected, 199-226, an amendment by Representative Tauscher to transfer all $15.0 million from RNEP (and $6.0 million from ACI) to conventional technology for use against hard and deeply buried targets. The Senate tabled, 56-41, an amendment by Senator Dorgan to bar use of funds for RNEP, and adopted an amendment by Senator Nelson (FL) to require congressional authorization to start engineering development or subsequent phases of RNEP.

In action on the FY2004 energy and water development appropriations bill, the House Appropriations Committee recommended reducing RNEP funds to $5.0 million, for reasons noted earlier in the "Legislative Actions in the FY2004 Budget Cycle" section under ACI. The House Appropriations Committee recommended having RNEP funds support penetration by conventional as well as nuclear weapons:

> The Committee directs that funding provided for the Robust Nuclear Earth Penetrator (RNEP) be used for research on the problem of deep earth penetration through hard or hardened surfaces, including modeling and simulation of the use of advanced materials, and varied trajectories and speeds. The Committee further directs that the National Nuclear Security Administration (NNSA) coordinate the RNEP research program with ongoing programs at the Department of Defense relating to research on earth penetration to maximize the dual-use applicability for both conventional and nuclear weapons.[30]

The House agreed to this bill without amending the Weapons Activities account, which includes RNEP.

The Senate Appropriations Committee recommended providing $15 million for RNEP. On September 16, as detailed under ACI, the Senate tabled an amendment by Senator Feinstein to eliminate all FY2004 funds requested for RNEP and adopted an amendment by Senator Jack Reed that barred use of funds provided by H.R. 2754 for phase 3 or 6.3 or beyond for advanced nuclear weapon concepts including RNEP.

---

[30] U.S. Congress. House. Committee on Appropriations. *Energy and Water Development Appropriations Bill, 2004.* H.Rept. 108-212, p. 145.

Conference Issues and Outcomes

RNEP was at issue in the Energy and Water Development Appropriations Act conference. Action on it was of concern to many because, of the four initiatives, RNEP is the one closest to deploying a weapon in response to NPR guidance. Non-RNEP ACI studies are in early stages, and enhanced test readiness and rescinding the low-yield ban are not weapons programs. In contrast, RNEP, if it proceeds, would involve conversion of an existing weapon. Conversion, under the current plan, would not involve nuclear testing, and would require a much smaller production effort than would production of complete new weapons. Even if the conference decided to proceed with the RNEP study, and if Congress later decided to proceed with the actual modification, deployment would not likely be quick; it took six years to develop the B61-11 and certify it for acceptance into the nuclear stockpile,[31] though lessons learned from that process might expedite deployment of RNEP.

The NDAA provided the funds requested for RNEP; conferees did not comment on it. THE EWDAA provided $7.5 million. Further,

> The conferees remind the Administration that none of the funds provided may be used for activities at the engineering development phases, phase 3 or 6.3, or beyond, in support of advanced nuclear weapons concepts, including the Robust Nuclear Earth Penetrator.[32]

The FY2005 Budget Cycle

The FY2005 request document seems to cast serious doubt on assertions that RNEP is only a study. Beginning with the FY2005 budget cycle, NNSA presented a detailed 4-year projection along with the current request; for RNEP, the figures are: FY2005, $27.6 million; FY2006, $95.0 million; FY2007, $145.4 million; FY2008, $128.4 million; and FY2009, $88.4 million, for a five-year total of $484.7 million.[33] Further, the document states the performance targets for RNEP as follows:

---

[31] "B61-11 Enters the Stockpile," *Weapons Insider* [published by the Nuclear Weapons Program of Los Alamos National Laboratory], September/October 2001: 1-2.

[32] Committee of Conference, Making Appropriations for Energy and Water Development for the Fiscal Year Ending September 30, 2004 ..., H.Rept. 108-357, p. 156.

[33] U.S. Department of Energy. Office of Management, Budget, and Administration/CFO. *FY 2005 Congressional Budget Request.* volume 1, National Nuclear Security Administration. DOE/ME-0032, February 2004, p. 63. The RNEP budget is available under "Directed Stockpile Work" at [http://www.mbe.doe.gov/budget/05budget/index.htm].

FY2005: "Complete 56% of scheduled RNEP Phase 6.2/6.2A activity." Further, "In FY2005, subsystem tests and a full system test of the proposed design will be completed."

FY2006: "Complete 100% of scheduled RNEP Phase 6.2/6.2A activity."

FY2007: "Report results of RNEP Phase 6.2/6.2A to Nuclear Weapons Council [a joint DOD-DOE agency that coordinates nuclear weapon requirements, production schedules, etc.] Obtain, if applicable, RNEP Phase 6.3 appropriate authorization. Complete initial 25% of scheduled RNEP Phase 6.3 activity (if authorized)."

FY2008: "Complete 65% of RNEP Phase 6.3 activity (if appropriately authorized)."

FY2009: "Complete 100% of scheduled RNEP Phase 6.3 activity (if authorized). Complete 15% of scheduled RNEP Phase 6.4 activity (if appropriately authorized)."[34]

An NNSA manager responsible for the program maintained that the budget increase beyond FY2005 is an artifact of the budget process. He stated that the money was inserted in the out years as a "placeholder" to protect the option of proceeding with RNEP. Were this not done, it is argued that NNSA would face two choices that it deems unsatisfactory: (1) By the time the budget for one fiscal year is submitted, the budget for the next fiscal year is largely fixed; without the placeholder, a decision to proceed with RNEP could therefore not be implemented until the second fiscal year. (2) Alternatively, without the placeholder, a decision to proceed with RNEP could be implemented promptly only by taking the needed funds out of other programs. Similarly, the move to Phases 6.3 and 6.4 reflects how the program might be expected to advance if it proceeds. The official, however, indicated that no decision has been made on whether or not to proceed with RNEP pending completion of the study.[35]

The RNEP study was initially projected to cost $45 million – $15 million a year for FY2003-FY2005. The numbers, however, have changed for each year. For FY2003, delay in submission of a DOD study required by the FY2003 National Defense Authorization Act (P.L. 107-314, Sec. 3146) delayed the start of NNSA's RNEP study; as a result, $6.0 million was spent of the $15.0 million appropriated. For FY2004, Congress cut the RNEP appropriation to $7.5 million. For FY2005, the request is $27.6 million, vs. the $15.0 million originally planned. Finally, FY2006, not FY2005, will be the last year of the RNEP study; NNSA estimates the FY2006 request at $30 million. The 4-year

[34] Department of Energy, *FY 2005 Congressional Budget Request,* volume 1, p. 69, 70, 76.

[35] Telephone interview, February 10, 2004.

total is about $71 million. Owing to the uncertainties of the program, NNSA could not, as of early March 2004, project an RNEP budget for FY2007-FY2009.

According to NNSA, the study's cost has grown for a number of reasons. The $45 million did not take into account participation in the study by Y-12 Plant, which would make components of RNEP, or of Pantex Plant, which would convert existing weapons into RNEPs; their participation adds some $2 million. DOE has imposed additional project management requirements that add $2 million. The rest of the increase comes from a better definition of the requirements of the study, refinement of cost estimates, and an increase in surety (safety, security, and use control) of the proposed weapon. On the latter point, DOE requires that any modifications of a nuclear weapon look for ways to increase its surety.[36] NNSA says it has found ways to increase RNEP surety, and plans to do so.

Critics are likely to view the outyear plan in the FY2005 budget request document as indicating an intent to proceed with RNEP. They may also question how the cost of the RNEP study grew some 58%. Congress, the media, and others, however, have not generally had access to the information presented here on the RNEP budget. This report will add views of critics and others when they become available.

ISSUES FOR CONGRESS

Would RNEP Promote Deterrence?

In the post-9/11 world, a new set of threats has come to the fore, including threats from rogue states seeking WMD. To deter rogue states, RNEP supporters maintain that the United States must hold at risk those assets that the leaders value most, and that in some instances RNEP may be the only way to do so. Everet Beckner, NNSA Deputy Administrator for Defense Programs, reportedly opined: "If we've got serious enemies in the world, I would like them to be deterred as much as possible by our military capabilities. ... I want them to be afraid of us."[37] Similarly, the House Policy Committee argued:

---

[36] U.S. Department of Energy. Order DOE O 452.1B, "Nuclear Explosive and Weapon Surety Program," approved August 6, 2001, Section 4(f).

[37] John Fleck, "Should Nuclear Research Resume?," *Albuquerque Journal,* March 17, 2003. [http://www.abqjournal.com/paperboy/ia/scitech/846058news03-17-03.htm]

> Credible deterrence requires that the President be able to hold at risk the things most important to an adversary who would seek to attack America. Deep underground facilities, including hardened bunkers and hard-rock tunnels, provide effective haven from attack. ...
> The President should have options — the options of conventional forces, of precision conventional weapons, and of nuclear weapons that are capable of holding all targets at risk.[38]

Critics question whether it is correct to apply the Cold War model of deterrence to rogue states, whose leaders might not be overly concerned about a U.S. nuclear attack on a few facilities in their nation or about the casualties it would cause. They might even expect to benefit from worldwide condemnation of the United States that an attack would likely generate. Furthermore, some RNEP critics argue that the use of any nuclear weapon would not be credible because of the massive damage and casualties that would result, and thus would not enhance deterrence.

Would RNEP Provide Added Military Value?

RNEP's supporters note that the Nuclear Weapons Council, a DOD-DOE organization chaired by DOD to coordinate nuclear weapons activities between the two agencies, approved the conduct of the RNEP study, and that DOD developed an Operational Requirements Document for RNEP that gives the desired specifications and capabilities.[39] They assert that RNEP is not redundant with the B61-11 because of the latter weapon's limitations.

Critics assert that nuclear EPWs are not needed. They note Brooks's statement, "we have no requirement to actually develop any new weapons at this time."[40] Representative Skelton stated, "the key to defeating hard deeply buried targets lies more in accuracy and penetration rather than the inherent explosive capability."[41] Many conventional options are available for defeating such targets.[42] U.S. forces demonstrated the ability of ground troops to attack tunnel complexes in Afghanistan and the ability of precision conventional ordnance to destroy underground bunkers in Iraq. It would be better, in this view, to spend

---

38 U.S. Congress. House. House Policy Committee. Subcommittee on National Security and Foreign Affairs. *Differentiation and Defense: An Agenda for the Nuclear Weapons Program.* February 2003, p. 7. [http://wilson.house.gov/media/photos/nuclearreport.pdf]

39 Information provided by NNSA staff, August 18, 2003.

40 Statement of Linton F. Brooks, April 8, 2003, p. 6.

41 U.S. Congress. *Congressional Record,* May 21, 2003: H4505.

42 Michael Levi, *Fire in the Hole: Nuclear and Non-Nuclear Options for Counterproliferation.* Non-Proliferation Project, Global Policy Program, Carnegie

funds on improving the ability to destroy these targets with conventional means rather than on nuclear weapons.

Would Fallout from RNEP Bar Use of the Weapon?

Critics of RNEP point to at least one claim by EPW supporters that low-yield EPWs would limit collateral damage sufficiently to contemplate their use.[43] (RNEP could well have significantly higher yields.) Critics challenge this view. They argue that no EPW could penetrate deeply enough to be fully contained, so that "the goal of a benign earth-penetrating nuclear weapon is physically impossible,"[44] as the descriptions of nuclear tests Des Moines and Baneberry make clear. For example, Nelson calculates

> A one kiloton earth-penetrating "mininuke" used in a typical third-world urban environment would spread a lethal dose of radioactive fallout over several square kilometers, resulting in tens of thousands of civilian fatalities.[45]

Under these conditions, it is argued, RNEP would be unusable:

> the vast resulting collateral damage to noncombatant populations would presumably limit employment of the weapon to retaliatory, or "intra-war" preemptive use in scenarios of all-out warfare, involving another nuclear weapon-state with the resources to both construct such deeply buried targets and threaten the survival of the U.S. as a nation-state — that is, China and Russia. ... the weapon is strategically, legally, and morally unsuitable for preemptive or retaliatory counter-proliferation warfare.[46]

Similarly, Representative Frank stated:

> ... we have said in Afghanistan, in Iraq, we are these days likely to be ... in an effort to rescue a people from an oppressive government. How welcome will

Endowment for International Peace, working paper no. 31, November 2002, p. 17-25.

43 Nelson, "Low-Yield Earth-Penetrating Nuclear Weapons," p. 2, references Dowler and Howard, "Countering the Threat of the Well-Armed Tyrant." The latter article, p. 36, argues that blast, crater radioactivity, and fallout would be confined to a small area. See also Robert Nelson, "Nuclear Bunker Busters, Mini-Nukes, and the US Nuclear Stockpile," *Physics Today,* November 2003: 32-37.

44 Nelson, "Low-Yield Earth-Penetrating Nuclear Weapons," p. 2.

45 Ibid., p. 1.

46 Christopher Paine with Thomas Cochran, Matthew McKinzie, and Robert Norris, Countering Proliferation, or Compounding It?: The Bush Administration's Quest for Earth-Penetrating and Low-Yield Nuclear Weapons, Natural Resources Defense Council, May 2003, p. vi. [http://www.nrdc.org/nuclear/bush/abb.pdf]

> our wagon be when it comes to nuclear arms? Do we tell the people of Afghanistan, do not worry, we will free you from the Taliban by using nuclear weapons within your country. ... I think you undercut the whole notion that America can be coming to the rescue of the victims of oppression.[47]

The argument over the extent of fallout RNEP would cause is, in the view of RNEP supporters, secondary to the key point, which, they contend, is that circumstances might call for the use of this weapon. A President always has various military options, and no President since 1945 has chosen to use nuclear weapons. Nonetheless, nuclear weapons might at some point be the only way to achieve a critical military objective, supporters argue. In that case, EPWs would be preferable in this view because they could achieve the desired result at a lower yield, and with less fallout, than surface-burst weapons. For some such uses, the B61-11 might suffice, but RNEP may be needed for others.

Would RNEP Have an Adverse Effect on Nuclear Nonproliferation?

On the issue of whether the United States should avoid researching EPWs in order to lead the rest of the world by example, Representative Thornberry argued:

> It is an interesting line of argument to say that we make the world safer when we tie our hands behind our back, that the problem is with the United States, and that if we would just set a good example, the Saddam Husseins and the Kim Jong Ils and even the Putins would fall right in line ... The problem is not American strength. ... Peace comes when America is strong and when America has additional options.[48]

The United States, in the view of RNEP's supporters, will not cause other nations to abandon WMD programs by abandoning RNEP. Nor, they suspect, would study, development, and deployment of RNEP cause nations to pursue WMD programs; they assert that deployment of the B61-11 had no such effect, so they doubt that a second EPW would, either.

Critics see grave proliferation risks arising from RNEP. They ask how the United States can preach to rogue states to restrain their nuclear programs while this nation is exploring new nuclear weapon capabilities. They see RNEP as

---

[47] Representative Barney Frank, colloquy on the Markey amendment to H.R. 4546, the Bob Stump National Defense Authorization Act for Fiscal Year 2003, *Congressional Record,* May 9, 2002: H2340.

[48] U.S. Congress. *Congressional Record,* May 21, 2003: H4506.

tailored for use against rogue states, and fear that that implies a U.S. willingness to use the weapon for that purpose. They foresee serious consequences from such use. As Representative Markey said, using one nuclear weapon "will destroy our moral and political credibility to end the spread of weapons of mass destruction, especially nuclear weapons."[49] In this view, it is inconsistent for the United States to fight Iraq because of WMD and then to develop WMD itself. Even studying RNEP, critics contend, would imply that the United States is considering its use, which could make nonnuclear weapon states accelerate their efforts to obtain nuclear weapons to deter U.S. attack on them, or to build new underground facilities at much greater depth.

Will Targets be Available for RNEP?

Fewer nations hold potential future targets for RNEP in early 2004 than was the case a year earlier. Any government that emerges in Iraq would in all probability be unable and unwilling to develop WMD for many years. Libya is in the process of giving up its WMD; indeed, President Bush praised Colonel Muammar Ghadafi of Libya for making "the right decision" in ending its WMD programs.[50]

Moving from the strategic to the tactical level, a successful attack on a deeply buried facility by an earth penetrator weapon would require precise targeting information. Yet events of the last year raise troubling questions on this score. Judging from press reports, U.S. intelligence failed to detect entire WMD programs in Libya, Iran, and North Korea, and detected chemical, biological, and nuclear weapon programs in Iraq that apparently did not exist. Given that record, critics may well question whether intelligence can be sufficiently precise and reliable to pinpoint underground bunkers to attack.

RNEP's supporters would note that Iran and North Korea pose a much more serious nuclear threat than did Iraq and Libya. Iran agreed in December 2003, under pressure from Britain, France, Germany, and Russia, to permit more intrusive inspections by the International Atomic Energy Agency (IAEA). It did so in order "to prove that the Iranian nuclear program is for civilian purposes," Gholam Reza Aghazadeh, Iran's Vice President, was quoted as saying.[51] The inspections, however, discovered equipment and materials that could be linked

---

[49] Ibid.

[50] U.S. White House. "President Announces New Measures to Counter the Threat of WMD." Remarks by the President on Weapons of Mass Destruction Proliferation, Fort Lesley J. McNair, National Defense University, Washington, DC, February 11, 2004.

[51] Christine Hauser and Nazila Fathi, "Iran Signs Pact Allowing Inspection of Its Nuclear

to nuclear weapons.[52] Shortly after these discoveries, Iran appeared to begin pulling back from full disclosure. A senior Iranian official reportedly said that his country "has given enough answers to the [IAEA's] questions. ... We have other research projects which we haven't announced to the agency and we don't think it is necessary to announce to the agency."[53] North Korea's statements have been contradictory and confusing for some time, but press reports have presented evidence that that nation has a program to produce highly enriched uranium, plutonium, and nuclear weapons, and that it received aid from Pakistan.[54]

Regarding the ability of tactical intelligence to provide the basis for nuclear weapon use, RNEP supporters would note that the bar against such use is extremely high, but that many intelligence assets, whether satellite imagery, communication intercepts, or information from defectors, would be brought to bear to ensure that the target was in fact connected with WMD.

Sites," *New York Times,* December 18, 2003.

52 See, for example, David Sanger, "In Face of Report, Iran Acknowledges Buying Nuclear Components," *New York Times,* February 23, 2004: 6; and Karl Vick, "Another Nuclear Program Found in Iran," *Washington Post,* February 24, 2004: 1.

53 "Iran Digs in Heels over Nuclear Secrets," Reuters, February 25,2004.

54 On the latter point, see David Rohde and David Sanger, "Key Pakistani Is Said to Admit Atom Transfers," *New York Times,* February 2, 2004: 1.

Chapter 6

NUCLEAR TEST READINESS

DESCRIPTION

During the Cold War, when the United States conducted nuclear tests on an ongoing basis, it took 18-24 months to prepare a typical test. With the end of the Cold War and the start of a moratorium on nuclear tests, the Clinton Administration required a U.S. ability to conduct a nuclear test in at most 24 to 36 months of a presidential order to conduct the test. DOD and NNSA, however, found that it would take close to 36 months to test, and they are uncertain as to their continued ability to test within even this length of time. Accordingly, they recommended "enhanced test readiness," or moving to and maintaining a posture in which tests could be conducted in less time. They supported an 18-month test readiness posture.

For FY2004, the Administration requested $24.9 million to maintain the existing posture and transition to the 18-month posture.[1] This figure is in addition to funds to maintain the Nevada Test Site, where NNSA would conduct any nuclear tests. NNSA indicated that it has tightly integrated the test readiness and enhanced test readiness programs into one program called "test readiness," so that it would be difficult to split them into the two components. Of the requested $24.9 million, however, it would take at least $15.0 million to maintain the current 24- to 36-month posture.[2] The NDAA provided the amount requested for test readiness; the EWDAA did as well, but conferees stated their

---

[1] U.S. Department of Energy. Office of Management, Budget and Evaluation/CFO. *Department of Energy FY 2004 Congressional Budget Request.* Volume 1, National Nuclear Security Administration. DOE/ME-0016, February 2003, p. 293.

[2] Personal communication, Kerry Webb, Manager, Test Readiness Program, NNSA, September 29, 2003.

expectation that NNSA would focus on restoring a 24-month posture before requesting "significant additional funds to pursue" an 18-month posture.

Consistent with the legislation, NNSA plans to move toward a 24-month readiness posture in FY2004 and achieve it in early FY2005. It plans to reach an 18-month posture later in FY2005, and requests $30.0 million for FY2005 for test readiness.

TECHNICAL BACKGROUND

Between 1951 and 1992, the United States conducted nuclear tests on a routine basis, with the numbers ranging from 96 in 1962, to 0 in 1959 and 1960 due to a moratorium, to 6 in 1992, the most recent year of U.S. testing, for a total of 1,030 tests.[3] As a result, the United States was always prepared to test, although "18 months was a minimal time to design and prepare most tests."[4] All U.S. tests since 1963 have been held underground, almost all at the Nevada Test Site.

In October 1992, the United States began a moratorium on nuclear testing that continues to the present. Ambassador Linton Brooks, NNSA Administrator, said in December 2003, "The policy of the United States is that there is not now a need for testing." He further stated that he did not see a reason to resume testing in the next two or three years.[5]

During the era of testing, DOE and its predecessors conducted several types of nuclear tests, each with its own goals, schedules, and characteristics. Some are explained here; it is useful to understand the differences in order to place test readiness in context.

Weapons modification and development. New-design warheads, i.e., those with a new primary or secondary stage, typically required nuclear testing. Most U.S. nuclear tests were for weapons development. In addition, nuclear testing was occasionally used to understand a problem in a stockpiled warhead, or to check a fix made to correct the problem, or for both purposes.[6] To conduct such

---

[3] Department of Energy, *United States Nuclear Tests, July 1945 Through September 1992,* p. viii. In addition to the 1,030 U.S. tests, 24 joint U.S.-U.K. tests were held at the Nevada Test Site

[4] U.S. Department of Energy. National Nuclear Security Administration. *Report to Congress: Nuclear Test Readiness,* April 2003, p. 6.

[5] "The Bush Administration's Views on the Future of Nuclear Weapons: An Interview with NNSA Administrator Linton Brooks," *Arms Control Today,* January/February 2004: 4, 6, 7.

[6] Information provided by Livermore staff, September 17, 2003. A Livermore study states, "Fifteen U.S. nuclear weapon systems have required post-deployment nuclear testing to identify or resolve problems." George Miller, Paul Brown, and Carol Alonso. *Report to*

tests, the labs had to design the test, fabricate a one-of-a-kind test device, fabricate the "diagnostics" (the elaborate set of instruments needed to gather the data), and emplace and seal the device and diagnostics in a shaft. These tests typically required 18 to 24 months,[7] counting from when the test was placed on the test calendar until it was conducted.[8] Most tests used standard diagnostics; designing non-standard diagnostics added time. Developing a new warhead typically required several tests; some fixes required one test, others needed several. Most modification and development tests were conducted in vertical shafts in the ground; drilling them typically took 6 to 8 weeks.[9]

Effects tests. Other tests examined how a nuclear explosion would affect military equipment, materials, electronics, nuclear warheads, etc., by exposing them to radiation from a nuclear explosion. These tests typically used a low-yield nuclear device designed to be a reasonably predictable radiation source; in these cases, design and instrumentation of the device itself were not at issue. The experiments required elaborate preparation, such as creating a vacuum chamber to test how x-rays would affect satellites. Effects tests typically used horizontal tunnels and experimental chambers bored into a mountain; excavation of the site for one such test, using three shifts a day, might take a year.[10] Planning the excavation, emplacing the experiments, and sealing the tunnels took added time. Such tests were "inherently a two to three year process."[11] A small fraction of U.S. tests were of this type.

Stockpile confidence tests. Beginning in the 1970s, DOE would pull a warhead from the field (e.g., a bomb deployed at a bomber base), modify it minimally as needed to make certain that its yield would not exceed the 150-kt ceiling of the Threshold Test Ban Treaty and to make it compatible with testing procedures, and test it. The purpose was to assure that weapons produced in quantity and then deployed — as distinct from one-of-a-kind devices built for weapons development tests — had the expected yield and met other basic performance criteria. These tests typically used diagnostics comparable to weapon development tests but less complex than for effects tests. They required

Congress on Stockpile Reliability, Weapon Remanufacture, and the Role of Nuclear Testing, report UCRL-53822, Lawrence Livermore National Laboratory, October 1987, p. 19. Note that many of the 15 were older systems.

7 U.S. Department of Energy. National Nuclear Security Administration. Nevada Operations Office. *Enhanced Test Readiness Cost Study,* July 1, 2002, DOE/NV-828, p. 2, 10.

8 Information provided by NNSA staff, September 11, 2003.

9 Ibid., p. 18.

10 Office of Technology Assessment, The Containment of Underground Nuclear Explosions, p. 20.

11 Department of Energy, Enhanced Test Readiness Cost Study, p. 21.

less time for preparation than weapon development tests, typically about a year. They were conducted in vertical shafts, a little less than once a year.[12]

Tests intended as political statements. When the Soviet Union terminated the 1958-1961 nuclear test moratorium in September 1961, the United States responded with a series of tests beginning two weeks later,[13] even though, by one analysis, "The United States was not prepared to conduct major experiments!"[14] All these tests had yields below 20 kt.[15] Such tests might be done in the future to respond to a nuclear test by Russia or China. They would surely use existing instruments and nuclear devices, and would probably involve experiments planned in advance to maximize their technical value. NNSA finds that it may be feasible to conduct a test of this sort in as little as 6 months, though that posture would entail added costs compared to an 18-month posture, would divert personnel from other tasks, and would require considerable advance preparation.[16]

At present, the only nuclear-related tests that NNSA is conducting are "subcritical experiments" (SCEs). CRS offers the following definition based on documents and on discussions with DOE and laboratory staff: "Subcritical experiments involve chemical high explosives and fissile materials in configurations and quantities such that no self-sustaining nuclear fission chain reaction can result. In these experiments, the chemical high explosives are used to generate high pressures that are applied to the fissile materials." Twenty SCEs were held between July 1997 and September 2003. SCEs study the behavior of plutonium.[17] One common purpose is to determine if material changes induced by radioactive decay of aged plutonium would degrade weapon performance. All have been held in a tunnel complex about 1,000 feet underground at Nevada Test Site. The United States sees them as consistent with the moratorium

---

[12] Information in this paragraph and note provided by Lawrence Livermore National Laboratory staff, September 10 and 17, 2003. The first stockpile confidence test (SCT) using a format agreed to by DOD and DOE was held in 1979. Another four tests — not called SCTs but that were for all practical purposes — were held beginning in 1972 before the first SCT. Counting these four tests, DOE conducted a total of 17 SCTs.

[13] Department of Energy, United States Nuclear Tests, July 1945 Through September 1992, p. 10.

[14] Harold Karan Jacobson and Eric Stein, Diplomats, Scientists, and Politicians: The United States and the Nuclear Test Ban Negotiations, Ann Arbor, University of Michigan Press, 1966, p. 283.

[15] Department of Energy, United States Nuclear Tests, July 1945 Through September 1992, p. ii, 10-12.

[16] Department of Energy, *Nuclear Test Readiness,* p. 8-9.

[17] See ibid., p. 15, regarding the first 19 tests. Regarding the 20th test, see U.S. Department of Energy. Nevada Site Office. "National Nuclear Security Administration Scientists to Conduct Piano Subcritical Experiment," news release, September 16, 2003, 1 p. Available at [http://www.nv.doe.gov/news&pubs/newsreleases/Pdfs/Piano_NR09162003.pdf].

because they produce no fission chain reaction. They are also of value for maintaining the test site and exercising testrelated skills.

NNSA plans to conduct three SCEs in FY2004 and two or more in FY2005.[18] One of the FY2004 experiments, named "Armando," and two FY2005 tests, "Unicorn" and "Krakatau," will be more complex than many of the earlier experiments. Unicorn, for example, will be the first conducted in a vertical shaft dug from the surface. According to NNSA,

> Initial site preparation for this experiment is underway. This activity, and the means for emplacement of the experimental hardware into the vertical hole, will appear visually similar to those employed in underground nuclear tests conducted prior to the 1992 moratorium. ... In addition to providing important information for stockpile stewardship, the Unicorn subcritical experiment will exercise key NTS capabilities not otherwise exercised in experiments carried out at the U1a complex.[19]

Some of the smaller, simpler SCEs held at the U1a complex may be replaced by the Joint Actinide Shock Physics Experimental Research Facility, a gas gun that accelerates a projectile to high speed and sends it into a mass of plutonium.[20] As a result, the schedule of subcritical experiments for FY2005 and beyond is uncertain.[21]

In the future; the most probable type of nuclear test (as distinct from SCEs) would be to resolve a defect that emerged in a stockpiled weapon to remedy a safety or reliability problem. Such tests would probably be similar in form to weapon modification or development tests. Regardless of purpose, any future tests would be the first done in conjunction with the stockpile stewardship program, which came into existence after the moratorium began. Preparation for testing would involve three parallel tracks:

(1) (1): Technical work involving the nuclear explosive. To fix a warhead problem, NNSA would "assess the problem, develop and implement a solution, and plan and execute a test that would provide the precise

[18] Information provided by NNSA staff, February 26, 2004.

[19] U.S. Department of Energy. National Nuclear Security Administration. Nevada Site Office. "'Unicorn' Subcritical Experiment Planned," news release, August 23, 2003, 1 p. Available at [http://www.nv.doe.gov/news%26pubs/newsreleases/Pdfs/UnicornSubcritical_NR08222003.pdf].

[20] For more detail, see U.S. Department of Energy. National Nuclear Security Administration. Nevada Site Office. "Nevada Test Site: JASPER." DOE/NV-709, rev. 1, February 2003, 2 p.

[21] Information provided by NNSA staff, February 26, 2004.

information needed to certify the 'fix.'"[22] NNSA estimates that it would take around 18 months to prepare this type of test.[23] As discussed below, some question this estimate. Technical preparation for other types of tests would take different lengths of time. Tests to make a political statement could be done in less than 18 months if the needed support activities were in place, and effects tests would take more than 18 months.

(2) (2): Support activities that can be done in advance to prepare for a generic test, such as conducting analyses of test safety, training personnel who can prepare the diagnostic equipment for a test, updating test-related agreements with federal and state agencies to ensure compliance with federal regulations, and obtaining or locating heavy equipment needed to prepare for a test.[24]

(3) (3): Support activities for a specific test, such as designing and building instruments to acquire the needed data, and analyzing the safety of the particular test in its emplacement site.

Test readiness involves bringing the generic steps to the point where completing them and support activities for a specific test would take a specified length of time. An NNSA source indicated that the need to spend 18 months to complete the first track is what led NNSA to select 18 months as the test readiness goal.[25]

History Through the FY2003 Budget Cycle

When the 1992 moratorium started, it was unclear how long it would last. The legislation creating the moratorium permitted it to end no sooner than July 1, 1993, but President Clinton continued it several times in the next several years. At the same time, he considered the possibility of a return to testing. His decision in August 1995 to press for a true zero-yield CTBT (as opposed to a treaty permitting nuclear tests of very low yield) was conditioned on several "safeguards," actions that the United States would take consistent with the treaty in order to protect U.S. interests despite the loss of testing. Safeguard F was:

---

[22] Department of Energy, *Nuclear Test Readiness,* p. 1.
[23] Ibid.
[24] Ibid., p. 7.
[25] Information provided by NNSA staff, February 27, 2004.

> The understanding that if the President of the United States is informed by the Secretary of Defense and the Secretary of Energy ... that a high level of confidence in the safety or reliability of a nuclear weapon type which the two Secretaries consider to be critical to our nuclear deterrent could no longer be certified, the President, in consultation with Congress, would be prepared to withdraw from the CTBT under the standard "supreme national interests" clause in order to conduct whatever testing might be required.[26]

As a result, it was unclear how ready the nuclear weapons laboratories and the Nevada Test Site should be to conduct a nuclear test. Maintaining readiness to conduct a test promptly would increase cost at the test site and the labs, and require personnel to spend time on readiness rather than on ongoing tasks. In a November 1993 Presidential Decision Directive, President Clinton decided to maintain the ability to conduct a nuclear test in 24 to 36 months of a presidential order to test. This so-called test readiness posture was first established in FY1996.[27]

Congress, the Administration, and others have expressed concern over the length of time needed to test. In January 1996, the Senate passed, 87-4, the resolution of ratification of the START II Treaty. The resolution included a number of declarations expressing the sense of the Senate. Declaration 12(E) included the following language: "The United States is committed to maintaining the Nevada Test Site at a level in which the United States will be able to resume testing within one year following a national decision to do so."[28] In its FY2001 report, the Foster panel argued that the current posture of 24 to 36 months was "excessive" and that the United States should not be in a position where increasing test readiness was viewed as indicating a major problem with the nuclear arsenal. It noted some reluctance by NNSA and DOD to increasing test readiness because the test moratorium seemed unlikely to end, so that funds spent on test readiness would not be available for other projects. It recommended a test readiness posture of 3-12 months, and having devices for nuclear testing available on short notice.[29] The purportedly leaked December 2001 Nuclear Posture Review asserted that "the current 2-3 year test readiness posture will not be sustainable as more and more experienced test personnel

---

[26] U.S. White House. Office of the Press Secretary. "Fact Sheet: Comprehensive Test Ban Treaty Safeguards," 1 p., August 11, 1995.

[27] Department of Energy. *Nuclear Test Readiness,* p. 4.

[28] U.S. Congress. *Congressional Record,* "Treaty with the Russian Federation on Further Reduction and Limitation of Strategic Offensive Arms (The START II Treaty)," resolution of ratification, p. S461.

[29] John Foster, Chairman, et al. FY2001 Report of the Panel to Assess the Reliability, Safety, and Security of the United States Nuclear Stockpile. March 2002, p. 28-29.

retire" and "the 2-3 year posture may be too long to address any serious defect that might be discovered."[30]

There was some uncertainty over the state of test readiness. In October 1999, Secretary of Defense William Cohen, in a statement before the Senate Armed Services Committee, maintained, "We would be able to conduct a nuclear test within 18 months to 2 years of a decision to do so."[31] The next day, Secretary of Energy Bill Richardson testified, "The test site is up and running and ready — while we are capable of fielding a well instrumented test in 18 to 24 months, my scientists tell me we, if pressed, could conduct a simple test in less than 1 year."[32]

DOD and NNSA conducted several studies between 1999 and 2003 to determine the time needed to conduct a nuclear test and what the optimum posture should be. Based on these studies,

> NNSA concluded that because of a loss of expertise and degradation of some specific capabilities, the U.S. would more likely require about 36 months to test, with less confidence in being able to achieve the 24-month end of the range. Furthermore, as time passes without further action, the 36-month posture is viewed as increasingly at risk.[33]

While NNSA's Nevada Site Office, which manages the Nevada Test Site, received funds in 2001 and 2002 to buttress the 24- to 36-month test readiness posture, NNSA and DOD favored enhancing test readiness. In 2002, they decided to plan to move to an 18-month posture. The Nuclear Weapons Council endorsed this transition in September 2002, and on March 13, 2003, NNSA provided Congress a notification of intent to make this transition.[34] Congress provided $15 million, the requested amount for enhanced test readiness, in H.J.Res. 2 (P.L. 108-7), the Consolidated Appropriations Resolution for FY2003.[35]

In action on the FY2003 National Defense Authorization Act, P.L. 107-314, a provision in the House bill required the Secretary of Energy to report, with the FY2004 budget submission, on a plan and budget to enhance test readiness,

---

30 Nuclear Posture Review [Excerpts], p. 35-36.

31 U.S. Congress. Senate. Committee on Armed Services. *Comprehensive Test Ban Treaty.* Hearings, 106th Congress, 1st Session, [hearings held October 6 and 7, 1999], S. Hrg. 106-490, USGPO, 2000, p. 18.

32 Ibid., p. 109.

33 Department of Energy, *Nuclear Test Readiness,* p. 5.

34 Ibid., p. 1, 2.

35 U.S. Congress. Committee of Conference. *Making Further Continuing Appropriations for the Fiscal Year 2003, and for Other Purposes.* H.Rept. 108-10, 108th Congress, 1st Session, USGPO, 2003, p. 904.

specifically the time and budget for a one-year test readiness posture. The Senate bill had no similar provision. The conference bill contained a provision, Section 3142, as follows:

> SEC. 3142. Plans for Achieving Enhanced Readiness Posture for Resumption by the United States of Underground Nuclear Weapons Tests.
>
> (a) Plans Required. — The Secretary of Energy, in consultation with the Administrator for Nuclear Security, shall prepare plans for achieving, not later than one year after the date on which the plans are submitted under subsection (c), readiness postures of six months, 12 months, 18 months, and 24 months for resumption by the United States of underground nuclear weapons tests.
> (b) Readiness Posture. — For purposes of this section, a readiness posture of a specified number of months for resumption by the United States of underground nuclear weapons tests is achieved when the Department of Energy has the capability to resume such tests, if directed by the President to resume such tests, not later than the specified number of months after the date on which the President so directs.
> (c) Report. — The Secretary shall include with the budget justification materials submitted to Congress in support of the Department of Energy budget for fiscal year 2004 (as submitted with the budget of the President under section 1105(a) of title 31, United States Code) a report on the plans required by subsection (a). The report shall include —
>
> (1) an assessment of the current readiness posture for resumption by the United States of underground nuclear weapons tests;
> (2) the plans required by subsection (a) and, for each such plan, the estimated cost for implementing such plan and an estimate of the annual cost of maintaining the readiness posture to which the plan relates; and
> (3) the recommendation of the Secretary, developed in consultation with the Secretary of Defense, as to the optimal readiness posture for resumption by the United States of underground nuclear weapons tests, including the basis for that recommendation.

The conference report further "[encouraged] the Secretary of Energy to submit plans for achieving and the cost of achieving and maintaining the recommended test readiness posture with, or as part of, the report required by the provision."[36]

The required report, with a cover date of April 2003, recommended a test readiness posture of 18 months; estimated that achieving this posture by the end of FY2005 would cost $83 million for the three-year transition period and that sustaining it would cost $25-$30 million a year thereafter for resources unique to nuclear testing, vs. about $15 million a year for this purpose for the current

[36] U.S. Congress. Committee of Conference. *National Defense Authorization Act for Fiscal Year 2003.* H.Rept. 107-772, 107th Congress, 2nd Session, USGPO, 2002, p. 786.

posture; assessed that the erosion of expertise over time due to retirements and the long time since the last nuclear test make the 24-36 month posture more likely a 36-month posture, with even that "viewed as increasingly at risk"; and argued that a 6-12 month posture would require "a substantial diversion of personnel and facilities at the laboratories and production plants," and at the shorter end of that posture "would represent a major redirection of the stockpile stewardship program."[37] NNSA also noted that it costs some $225 million a year to support general requirements at the Nevada Test Site such as infrastructure, personnel, and equipment.

Legislative Actions in the FY2004 Budget Cycle

In S. 1050, Section 3132 required the Secretary of Energy to achieve an 18-month nuclear test readiness posture by October 1, 2006, and to maintain that posture thereafter. If as a result of the review for the report required by P.L. 107-314, Section 3142, the Secretary, in consultation with the Administrator for Nuclear Security (the head of NNSA), determined that a posture other than 18 months was preferable, the Secretary "may, and is encouraged to, achieve and thereafter maintain" that posture, and shall submit a report stating the preferred posture and the reasons for it. NNSA's report on test readiness, however, indicated a preference for 18 months. In H.R. 1588, Section 3116 permitted DOE to obligate not more than 40 percent of FY2004 funds for nuclear test readiness until 30 days after DOE submitted the report required by P.L. 107-314, Section 3142. NNSA submitted that report, dated April 2003. There were no amendments to these provisions, or any others on enhanced test readiness, in the House or Senate.

The House Appropriations Committee recommended eliminating the $24.9 million requested for "enhanced test readiness" on grounds of inadequate budget justification and dubious program credibility. It was "concerned with the open-ended commitment to increase significantly funding for the purpose of Enhanced Test Readiness without any budget analysis or program plan to evaluate the efficiency or effectiveness of this funding increase."[38] It questioned NNSA's ability to maintain an 18-month readiness posture during a test moratorium when that was the minimum time to prepare a test when testing was ongoing. It linked budget and program issues: "the Committee will not spend money on a perceived problem when the Department [of Energy] has not

---

[37] Department of Energy, *Nuclear Test Readiness,* p. 1, 2, 5, 6.

[38] U.S. Congress. House. Committee on Appropriations. *Energy and Water Development Appropriations Bill, 2004.* H.Rept. 108-212, p. 148-149.

provided a rationale or a plan that addresses the underlying problems inherent in maintaining a testing capability during a testing moratorium."[39] The House passed H.R. 2754 as reported from the committee without amending the Weapons Activities section, which includes test readiness.

The Senate Appropriations Committee recommended providing the full $24.9 million requested, but did not comment on the readiness program. On September 16, the Senate rejected an amendment by Senator Feinstein on various nuclear programs, as detailed under ACI. One provision of the amendment would have barred use of funds in the bill for modifying the test readiness posture of the Nevada Test Site to a readiness posture of less than 24 months.

Conference Issues and Outcomes

With the House having eliminated the $24.9 million requested for test readiness and the Senate having provided the full request, the appropriation amount was at issue. Complicating a decision, the House eliminated $24.9 million for "enhanced test readiness," while the Administration requested $24.9 million for "test readiness." Some of the requested funds were to maintain the 24- to 36-month posture and some were to transition to an 18-month posture. Yet the House Appropriations Committee expressed concern over the latter even as it "supports the continued maintenance of the Nevada Test Site."[40] Eliminating the $24.9 million would have eliminated funds for both postures. If conferees decided to reduce funds for test readiness, would they have eliminated the full request, as the House voted to do, or eliminated only the funds for moving to an 18-month posture (perhaps $9.9 million), or settled on some other amount? (Note that the Nevada Test Site receives other funds. For example, the budget category "Nevada Site Readiness," for which DOE requested $39.6 million for FY2004, "[i]ncludes activities required to maintain the Nevada Test Site (NTS) that are not unique to the test readiness mission, but do support the stockpile stewardship mission."[41])

DOE's submission of the test readiness report rendered moot the limit on obligations in the House defense authorization bill and the provision in the Senate defense authorization bill regarding a test readiness posture of other than 18 months. Those provisions were not at issue in conference. While not a matter in dispute between House and Senate, authorization conferees could have

[39] Ibid., p. 149.
[40] Ibid., p. 148-150.
[41] Department of Energy, FT 2004 Congressional Budget Request, p. 291.

considered whether an 18-month test readiness posture should be achieved by October 1, 2006, as per the Senate bill, or by the end of FY2005, the date to which DOD and NNSA agreed. They could also have followed the House Appropriations Committee's lead and required DOE to provide a more comprehensive justification before endorsing any approach.

The NDAA conference report did not address test readiness. The EWDAA conference report provided the full funds requested, but changed the test readiness posture:

> The conferees recognize that test readiness activities in Nevada were allowed to atrophy during the last decade under the current nuclear test moratorium ...However, the conferees expect the NNSA to focus on restoring a rigorous test readiness program that is capable of meeting the current 24-month requirement before requesting significant additional funds to pursue a more aggressive goal of an 18-month readiness posture.[42]

Thus the conferees, by basing their decision on the need for NNSA to restore a 24-month posture before proceeding to an 18-month posture, simultaneously avoided the issue of what aspects of test readiness would have been cut by reducing funds below the request; expressed their support for reversing the "atrophy" of the test program; pressed NNSA to "restor[e] a rigorous test readiness program"; implied that it would be technically difficult to restore an 18-month posture; reflected concerns expressed by the House Appropriations Committee, Senator Feinstein, and others about moving to an 18-month posture; and left the door open to moving to that posture later.

THE FY2005 BUDGET CYCLE

The FY2005 request for test readiness is \$30.0 million.[43] As noted, conferees on the FY2004 energy and water development appropriations conference report "expect" NNSA to move to a 24-month posture before requesting "significant additional funds" to move to an 18-month posture. NNSA indicates that that provision does not restrict its actions on test readiness. It anticipates reaching a 24-month posture in early to mid FY2005, and hopes to achieve an 18-month posture by the end of FY2005.[44] Unlike ACI and RNEP, the budget document does not provide out-year budgets for test readiness.

---

42 Committee of Conference, Making Appropriations for Energy and Water Development for the Fiscal Year Ending September 30, 2004 ..., H.Rept. 108-357, p. 159-160.

43 Department of Energy, FY 2005 Congressional Budget Request, vol. 1, p. 86.

44 Telephone conversation with NNSA staff, February 26, 2004.

NNSA estimates, however, that it can maintain an 18-month readiness posture for $25 million a year once it has established that posture, which the FY2005 request would enable it to do.

ISSUES FOR CONGRESS

Is An 18-Month Test Readiness Posture Desirable?

Some see the current 24- to 36-month posture as adequate. In this view, enhanced test readiness would divert resources, as well as the time of experts, to a task that is unlikely to be carried out, to the detriment of the ongoing stockpile stewardship program that has so far enabled the United States to maintain its nuclear stockpile without testing. They believe that the current test readiness program has sufficient means to maintain skills, such as subcritical experiments, simulations, and exercises. They point to a National Academy of Sciences panel that stated:

> We judge that the United States has the technical capabilities to maintain confidence in the safety and reliability of its existing nuclear-weapon stockpile under the CTBT, provided that adequate resources are made available to the Department of Energy's (DOE) nuclear-weapon complex and are properly focused on this task.[45]

Some fear that enhancing test readiness could have adverse international consequences. One advocacy group was "alarmed that the Administration and Congress would take steps toward resuming nuclear testing, which would undo decades of arms control and set the stage for a new arms race."[46] Further, "U.S. resumption of full-scale nuclear testing would threaten the viability of the Nuclear Non-Proliferation Treaty." Representative Markey and 55 House colleagues wrote to the House Appropriations Committee that they "are concerned that the proposed enhanced test readiness program will renew interest in testing among other nations."[47]

---

[45] National Academy of Sciences. Committee on Technical Issues Related to Ratification of the Comprehensive Nuclear Test Ban Treaty. *Technical Issues Related to the Comprehensive Nuclear Test Ban Treaty.* Washington, National Academy Press, 2002, p.1.

[46] Alliance for Nuclear Accountability, letter to defense authorization conferees, September 5, 2002. [http://www.ananuclear.org/testreadylttr.html]

[47] Letter from Representative Edward Markey and 55 other Representatives to Honorable Sonny Callahan, Chairman, House Appropriations Committee, Energy and Water Development [Subcommittee], and Honorable Peter J. Visclosky, Ranking Member, House Appropriations Committee, Energy and Water Development [Subcommittee], July 9, 2002,

Stewardship may not require an 18-month readiness posture, in the view of those opposed to shortening the posture, because test readiness could be improved on an ad hoc basis if needed. It would in all likelihood take months to convince the President and Congress that a test was needed. The interval between discovery of a problem and a political decision to test could be used for preliminary planning for the test, reducing the time between the decision to test and the conduct of the test.

The other main purpose for testing beside stockpile stewardship is development of new weapons. Those favoring the current readiness posture see an 18-month posture as irrelevant for that purpose. The idea that the United States can only respond to a particular threat by testing a new weapon in 18 months, and then producing and using it, is not credible in this view, given the massive U.S. conventional and nuclear forces in being.

Those who would enhance test readiness argue that the current posture is simply too long. Secretary of Energy Spencer Abraham said, "Should our scientists decide we cannot certify the reliability of our nuclear stockpile, we must be capable of conducting a nuclear test in a much shorter time frame than the current three years."[48] NNSA says that the 24- to 36-month posture has become closer to a 36-month posture, and even that is at risk as people formerly involved in the test program retire.[49] It is difficult to maintain skills that would be needed for a resumption of testing under a 24- to 36-month posture, it is argued, because that posture implies that test-related activities would only be a minor part of the workload for staff who would be responsible for conducting nuclear tests. A more active test readiness program would sharpen their skills and help recruit people with skills needed for testing. It may even prove infeasible to stabilize the posture at its present unsatisfactory level as skills will erode. In this view, the choice would be between continuing current trends and reversing them. They dismiss charges that enhancing test readiness would lead automatically to testing; a decision to test would be decided separately by the President and Congress.

Those who would improve test readiness divide into two groups. NNSA claims that 18 months is the optimal time:

> An 18-month posture is appropriate because this is the minimum time we would expect it would take, once a problem was identified, to assess the problem, develop and implement a solution, and plan and execute a test that would provide the information needed to certify the 'fix.' An 18-month posture

p. 2, on RNEP and test readiness. [http://www.house.gov/markey/iss_nonproliferation_ltr020718.pdf]

48 Spencer Abraham, "Facing a New Nuclear Reality," *Washington Post,* July 21, 2003: 21.

49 Department of Energy, *Nuclear Test Readiness,* p. 5.

> also reflects what is cost effective while continuing to support other stockpile stewardship missions.[50]

Others feel that 18 months is too long. As noted, the Senate called for a oneyear test readiness posture and the Foster panel recommended a 3-12 month posture.

Still others believe that changing the test readiness posture would have little practical significance because technical considerations determine the time needed to conduct various types of tests. A press report quotes Raymond Jeanloz, a member of the former NNSA Advisory Committee, on this point:

> The committee was told by the national nuclear laboratories that "the nation would be able to perform a test in 3 to 6 months" if the goal was simply to produce an explosion, he said. "From the labs' point of view, until they know why they would have to have a test to address some hypothetical technical problem, they don't know how long it would take them. So this whole business of a three-year, or a one-and-a-half year, or a half-year delay before they can test is incredibly artificial," he said.[51]

Is An 18-Month Posture Feasible for Stockpile Stewardship or Weapons Development Tests?

As discussed earlier, tests to meet a political need could be conducted rapidly, while some complex tests, such as effects tests, would probably take longer than 18 months even with enhanced test readiness. Judging from history, such tests would be rare. At issue would be the time to conduct the most likely types of tests, notably those to remedy a problem in an existing warhead discovered through the stockpile stewardship program, or those to design new warheads.

NNSA claims that an 18-month posture should be feasible.

At the time of an active underground test program, 18 months was a minimal time to design and prepare most tests, and is therefore a reasonable indicator of the time that would be required to prepare a device for testing if the details of the planned test are not known in advance.[52]

---

[50] Ibid., p. 6.

[51] David Ruppe, "Energy Department might not release disbanded committee's report," *GovExec.com,* August 13, 2003: [http://www.govexec.com/dailyfed/0803/081303gsn2.htm] The advisory committee was created pursuant to the Federal Advisory Council Act (5 USC App.), and was reportedly disbanded in June 2003; ibid.

[52] Department of Energy, *Nuclear Test Readiness,* p. 6.

Critics respond that this argument is illogical. If it took a minimum of 18 months to prepare a test when the test program was in full swing, they ask, how can the nuclear test program expect to establish that as a maximum after more than a decade without testing? In the intervening years, test expertise has eroded through disuse, many people formerly in the test program have shifted to other areas or have retired or died, fewer new staff have been trained with these skills than have left the test program, construction crews would have to be found and trained, new diagnostic instruments would have to be designed to replace technology from 1992. Further, NNSA would have to meet regulatory standards for environment, safety, and health added since 1992.[53] It may be possible to reach 18 months eventually, but given the uncertainties, costs, and personnel issues, it is uncertain when that could occur.

53 Ibid., p. 10.

Chapter 7

CONCLUDING OBSERVATIONS

Debate over the four nuclear initiatives will in all likelihood continue for some years. This report concludes by discussing some substantive and logical aspects of the debate that cut across the four initiatives. The topics below speak to the nature of the debate in 2003 and may bear on it in future years.

FOUR SMALL ISSUES, OR ONE LARGE ONE?

Supporters of the four initiatives argue that opponents are reading too much into the initiatives. They are, in this view, unexceptional elements of a nuclear weapons program, each with its own justification. Supporters argue: lifting the ban on low-yield R&D simply permits scientists to study the full range of weapon issues, as is the case for scientists in all other nuclear weapon states; ACI is for early-stage studies; and RNEP is just a study that at most could lead merely to adding a second type of earth penetrator weapon to the U.S. inventory. A three-year test readiness posture is much too long, and testing may be needed for other purposes, such as to check on a fix to a warhead type that developed a defect. As Secretary of Energy Spencer Abraham said,

> We are not planning to resume testing; nor are we improving test readiness in order to develop new nuclear weapons. In fact, we are not planning to develop any new nuclear weapons at all. Our goal is designed to explore the full range of weapons concepts that could offer a credible deterrence and response to new and emerging threats as well as allow us to continue to assess the reliability of our stockpile without testing.[1]

---

[1] Spencer Abraham, "Facing a New Nuclear Reality," *Washington Post,* July 21, 2003: 21.

Domestic and foreign critics see these measures as parts of a whole — a military policy with a lowered threshold against nuclear use. They contend: lifting the lowyield ban will enable pursuit of more-usable weapons; and, ACI will present the military with new weapon options. In turn, military requirements for these new weapons will lead to testing, and enhanced test readiness will permit expedited testing and deployment of these weapons. They view RNEP as redundant, since the United States has another nuclear penetrator (the B61-11), conventional penetrators, and other nonnuclear options, but believe RNEP will make nuclear use that much more likely by giving the United States yet another nuclear option. From this perspective, these initiatives are consistent with other policy decisions, such as the Administration's policy of preemption, its willingness to use "all of our options," its disregard for negative security assurances, and its refusal to pursue the CTBT. Critics see this policy as extremely insensitive to world opinion at best, and potentially dangerous.

NUCLEAR PREEMPTION AND USE: AMBIGUITIES AND UNCERTAINTIES

The Administration's policy statements leave a crucial uncertainty. In September 2002, *The National Security Strategy of the United States of America* stated, "To forestall or prevent ... [attacks with WMD] by our adversaries, the United States will, if necessary, act preemptively."[2] And in December 2002, the *National Strategy to Combat Weapons of Mass Destruction* stated, "The United States will continue to make clear that it reserves the right to respond with overwhelming force — including through resort to all of our options — to the use of WMD against the United States, our forces abroad, and friends and allies."[3] These two statements do not specifically declare that the United States will use nuclear weapons preemptively against possible WMD attacks. Nonetheless, critics take them together as a sign that the Administration is contemplating just such a course. Is the Administration willing to rule out preemptive use of nuclear weapons? It may prefer ambiguity on this issue, whether because it has not reached a decision or as a tactic to extend deterrence. But by removing ambiguity, declaring that it would not use nuclear weapons preemptively, the Administration could reduce fears worldwide about possible

---

[2] U.S. President. The National Security Strategy of the United States of America. September 2002, p. 15.

[3] U.S. White House. National Strategy to Combat Weapons of Mass Destruction. December 2002, p. 3.

nuclear war. At issue for the Administration is deciding whether the benefits of resolving this ambiguity outweigh the costs.

WILL THE NEW WEAPONS DETER?

The Administration, expressing concern about the terrorist WMD threat, discusses a purpose of new or modified nuclear weapons as being to counter WMD facilities. The Administration may wish to clarify how the nuclear initiatives — or any military means, for that matter — could deter or retaliate against a terrorist WMD threat. If the perpetrator is unknown, or is known but has no known address, what targets would be attacked, and would new weapons offer any advantage for such attacks? Are any assets of sufficient value to terrorists that holding them at risk might deter terrorist attacks? If so, would current U.S. nuclear and conventional forces suffice for that purpose? Could the United States perhaps deter terrorist use of WMD by threatening to overthrow (using conventional forces) regimes that support terrorism, or by threatening to destroy (using nuclear or other forces) facilities that these regimes see as critical?

There is a more fundamental deterrence issue. Deterrence depends on holding at risk assets of great value to the leadership. But while the United States was able to calculate what assets needed to be held at risk to deter the Soviet Union (and vice versa), the United States cannot simply assume that what deterred the Soviet Union would deter rogue states. The leaders of a nuclear-armed rogue state might calculate that the United States would not use nuclear weapons against them because of the risk of international opprobrium, or that U.S. nuclear weapons would not have a decisive effect on the regime because it could hide or bury its own WMD and the facilities of highest value to the leaders. They might be willing to accept the use of several nuclear weapons against their country, especially if they follow the U.S. debates and come to believe that the United States would only use a few low-yield weapons of limited effectiveness.

WILL RESEARCH LEAD TO TESTING, ACQUISITION, AND USE?

A key concern of opponents of nuclear testing is the prospect that ACI, RNEP, and low-yield weapons development could lead to nuclear tests. One could imagine various ways in which these programs could lead to tests. Lifting the ban on low-yield R&D, for example, might lead to ACI developing new types of low-yield warheads: earth penetrators, chemical and biological agent neutralizers, warheads with reduced neutron or enhanced gamma radiation output, and warheads with reduced electromagnetic pulse. Future military interest in these warheads would, in this view, lead inexorably to testing.

Yet there are plausible arguments that testing would not be needed. The B61-11 penetrator was converted from a B61-7 bomb without testing, and the RNEP conversion anticipates converting an existing B61 or B83 without testing. Those warheads are available in a number of yields; it may be that a lower-yield option exists that would eliminate the need to develop a new-design EPW. The Administration has enough confidence in nonnuclear interceptors to withdraw from the ABM Treaty in order to deploy a system using them. Nuclear weapons tailored to destroy chemical and biological agents may not be needed. The ability to destroy these agents is arguably much more dependent on precise intelligence than on weapon characteristics, some doubt the efficacy of nuclear weapons to destroy such targets, and some doubt that the advantages of avoiding contamination by chemical or biological agents outweigh the disadvantages of contaminating an area with radiation. Beyond that, whether enhancing nuclear test readiness will lead to testing depends on policy rather than technical judgments on future weapons R&D.

Some critics of the Administration's nuclear initiatives fear that studying nuclear weapon concepts or new weapons will lead to their use, but that result is far from inevitable, supporters reply. Nuclear weapons are an option available to the United States, but so are conventional forces, diplomacy, and economic leverage. The United States has had low-yield nuclear weapons continuously since the 1950s, and might have benefitted tactically from using them in the Korean War, Cuba, and the Vietnam War. Yet despite these military benefits and options, Presidents refrained from using these weapons. It appears, then, that nuclear weapons would only be used as a last resort. On the other hand, any new weapon would provide a President with a wider range of options for nuclear use. A President might find options made available by current nuclear weapons to be unacceptable, but a new weapon might tilt the tradeoff between costs and benefits in favor of nuclear use.

Any New Weapon Has Limits to its Military Value

The claim that EPWs may be the only way to destroy certain hardened and deeply buried targets does not mean that any nuclear EPW will destroy any such target. Rather, nuclear EPWs would make a difference only against some specified range of targets. A bunker hardened to a given level and buried to depth X could be destroyed by a conventional EPW, and buried more deeply at depth Y could not be destroyed by a nuclear EPW of 5 kilotons yield. Only if the facility is buried between depths X and Y would that EPW make a difference. The United States could increase the weapon's yield, permitting it to destroy the bunker at greater depth, but the target nation could counter by increasing the hardness of the bunker, burying it deeper, or camouflaging its location. Further, the U.S. intelligence may not know details of the geology above the bunker or of the layout of a tunnel complex, both of which may greatly affect the weapon's effectiveness. Thus the weapon adds military value over only a specific range of targets.

INDEX

R

S

U

W